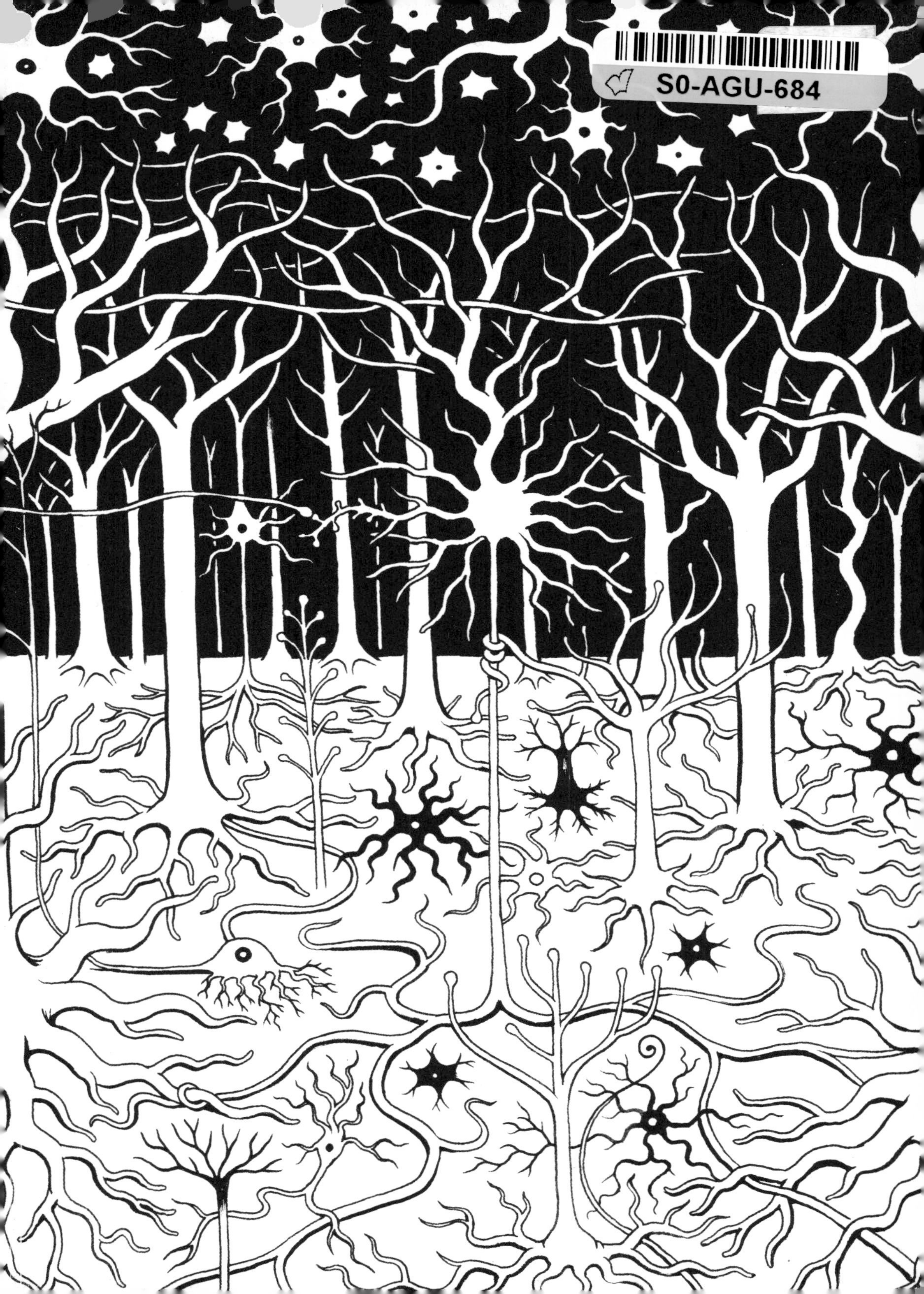

DR. MATTEO
FARINELLA

DR. HANA
ROŠ

NEUROCOMIC

Supported by
wellcometrust

NOBROW
PRESS

MIX
Paper from responsible sources
FSC® C101807

Published by Nobrow Ltd.
62 Great Eastern Street, London, EC2A 3QR

This is a first edition printed 2013.

Supported by wellcome trust

Printed in Belgium on FSC assured paper.

ISBN: 978-1-907704-70-3

Order from www.nobrow.net

PROLOGUE

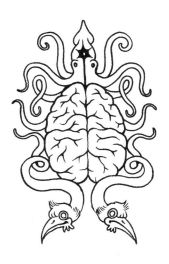

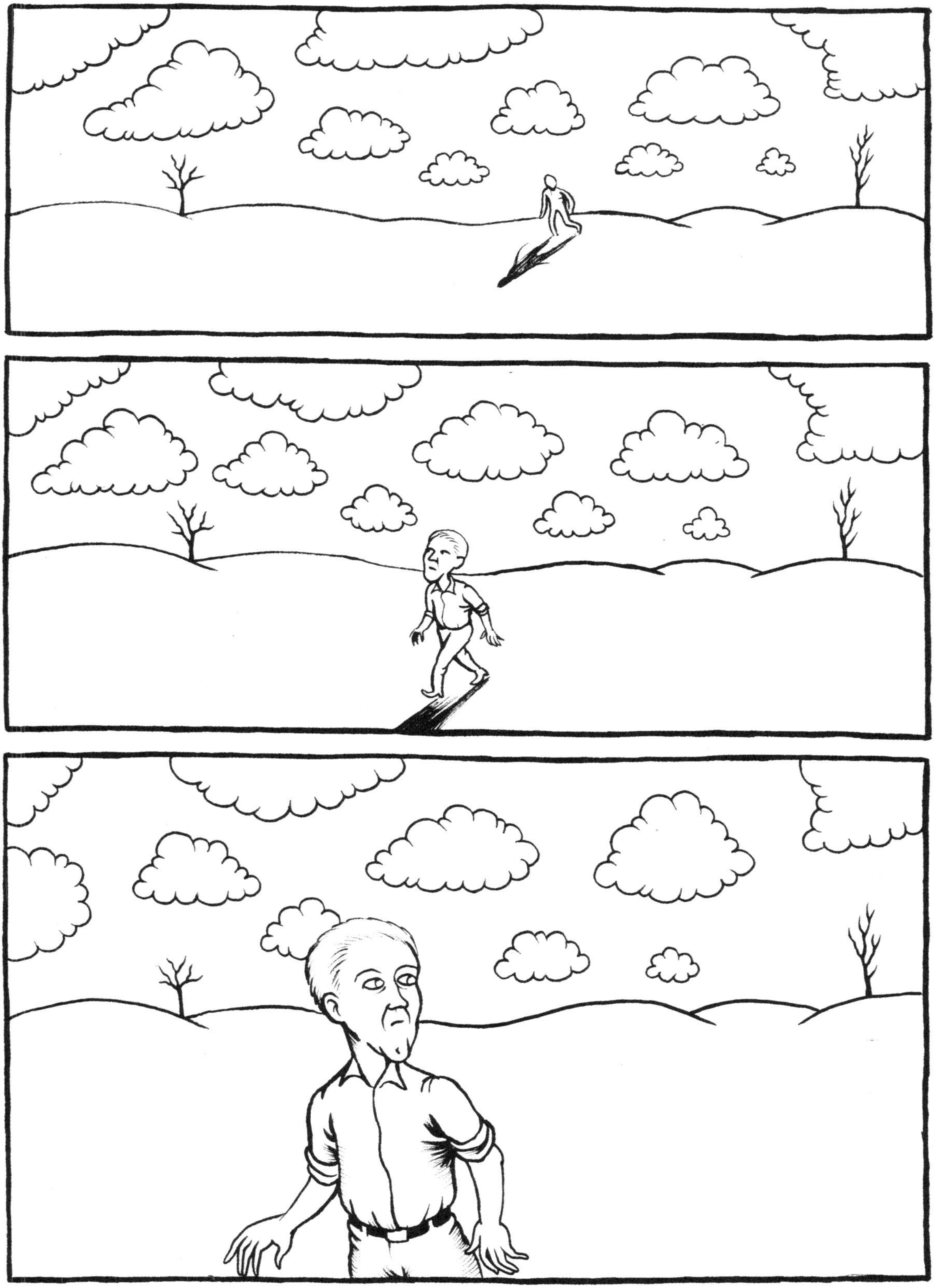

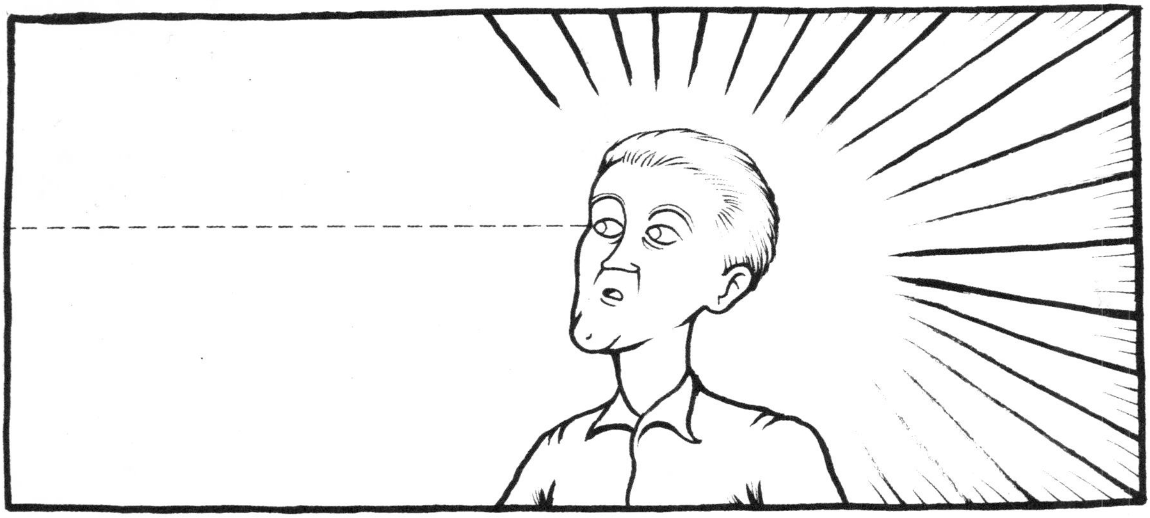

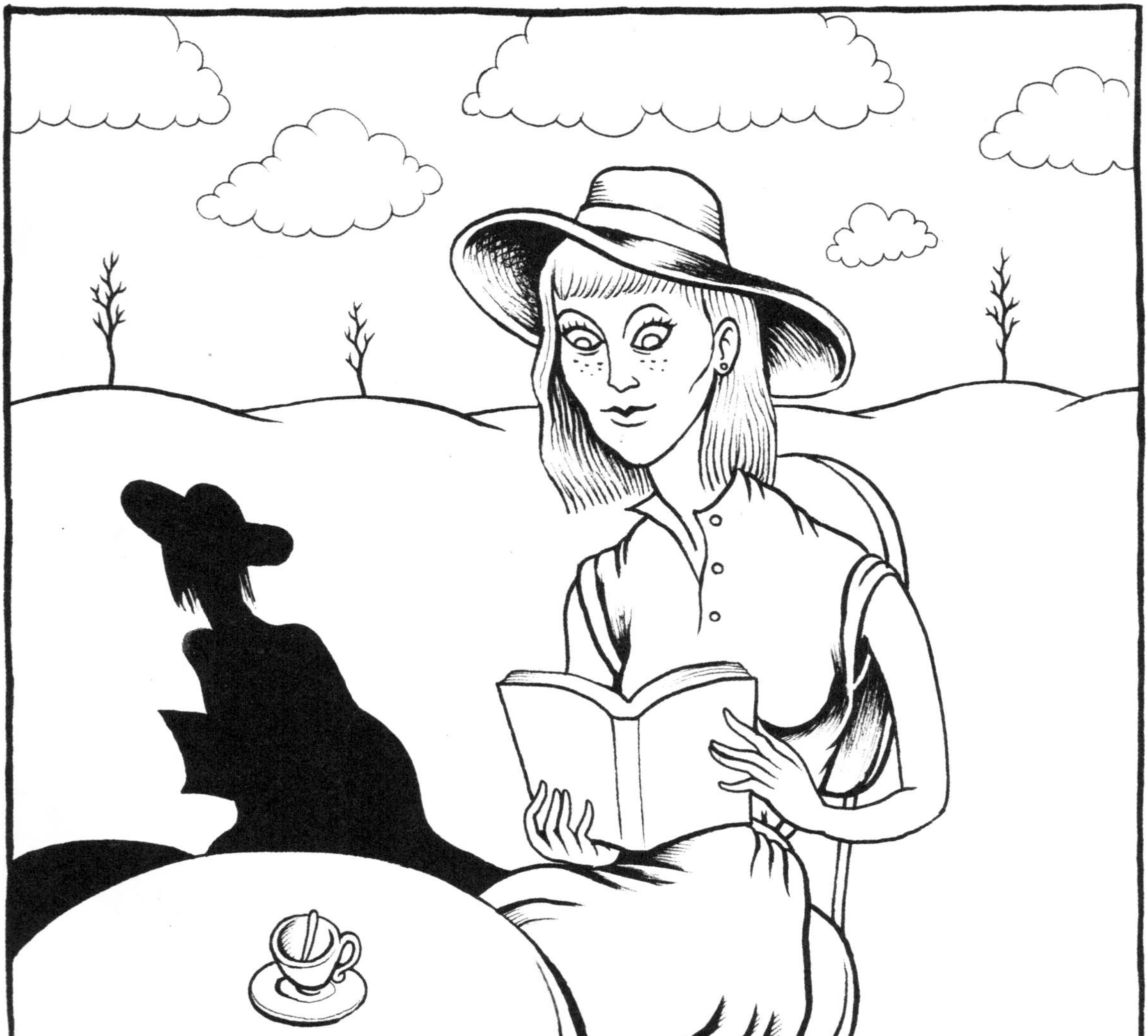

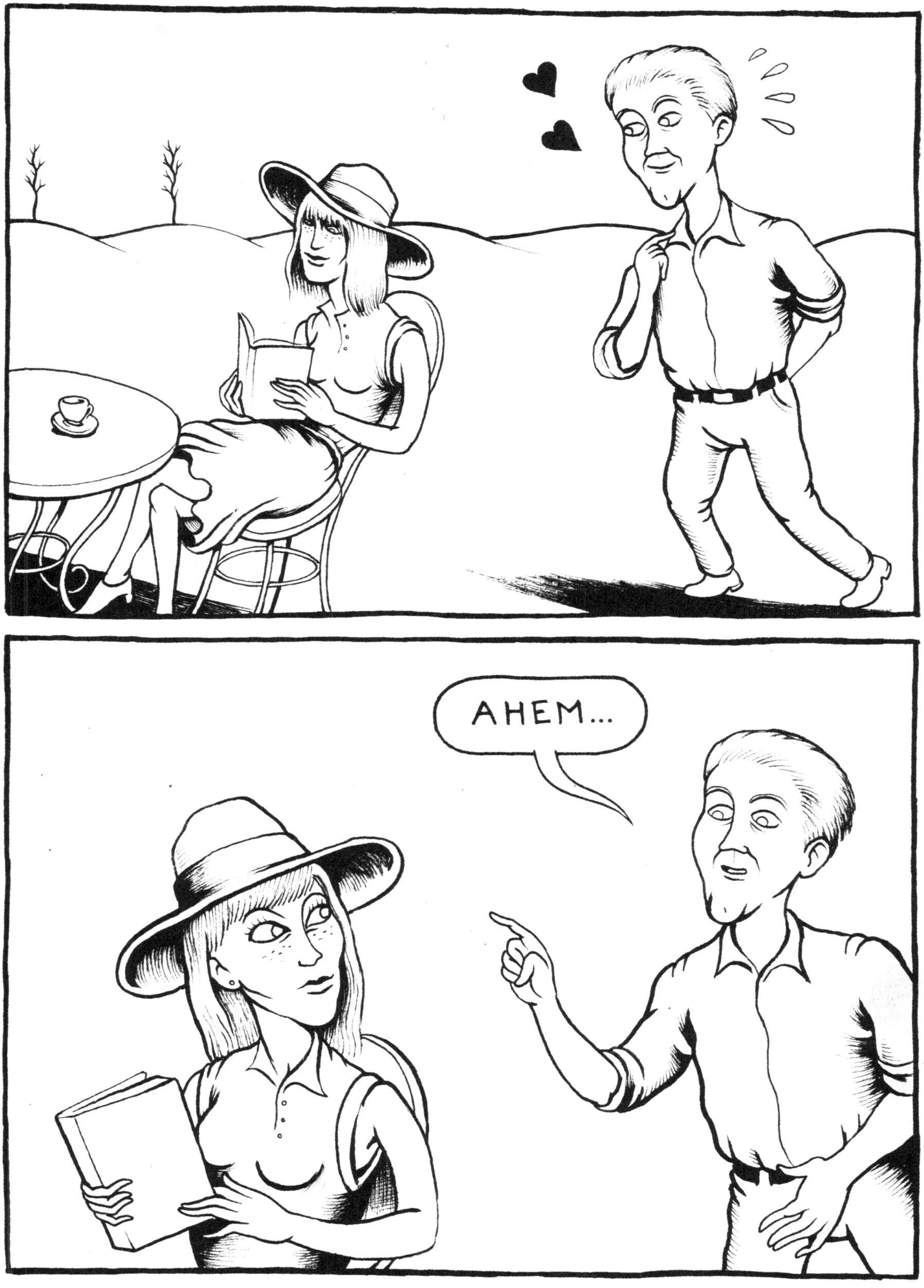
AHEM...

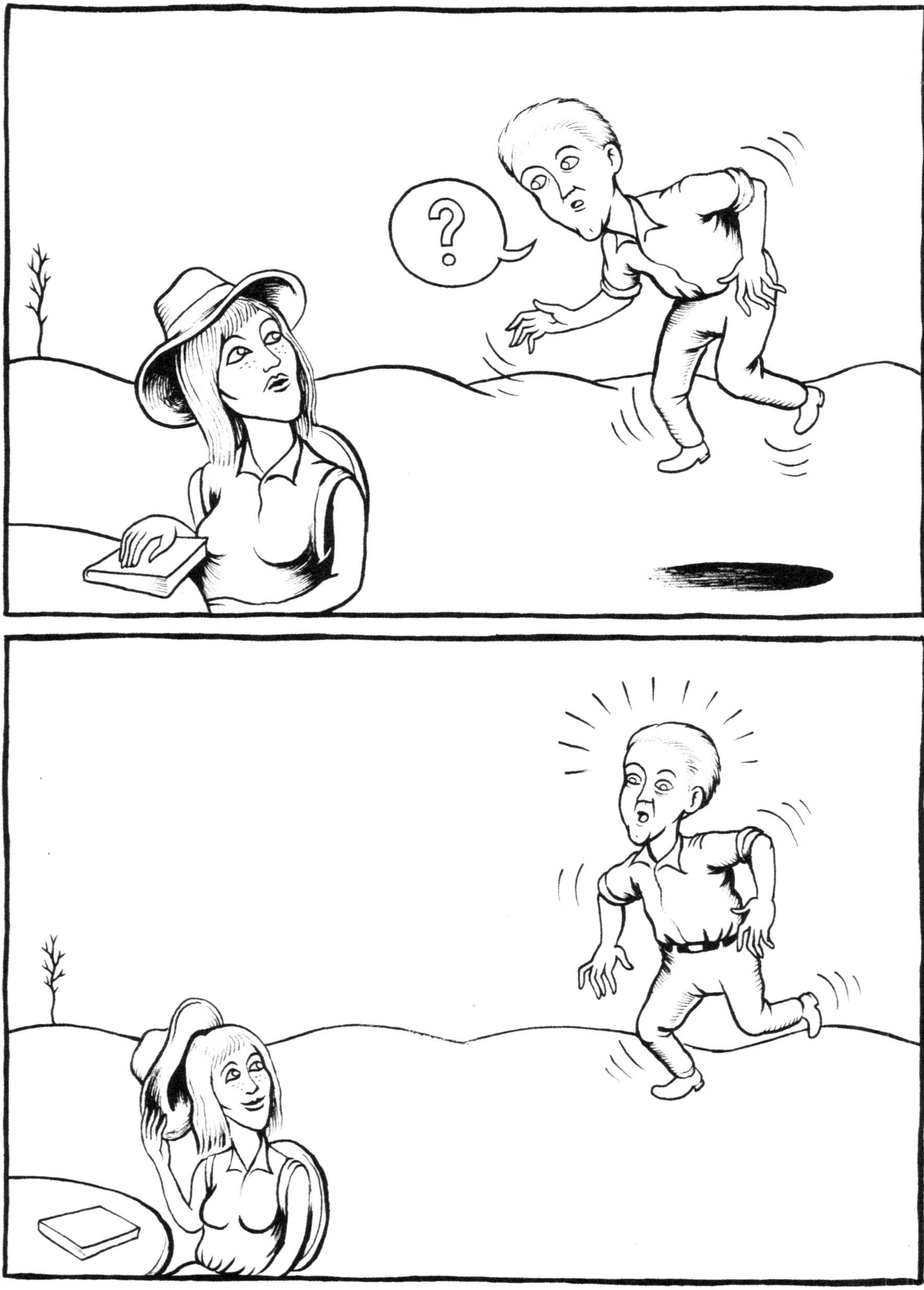
?

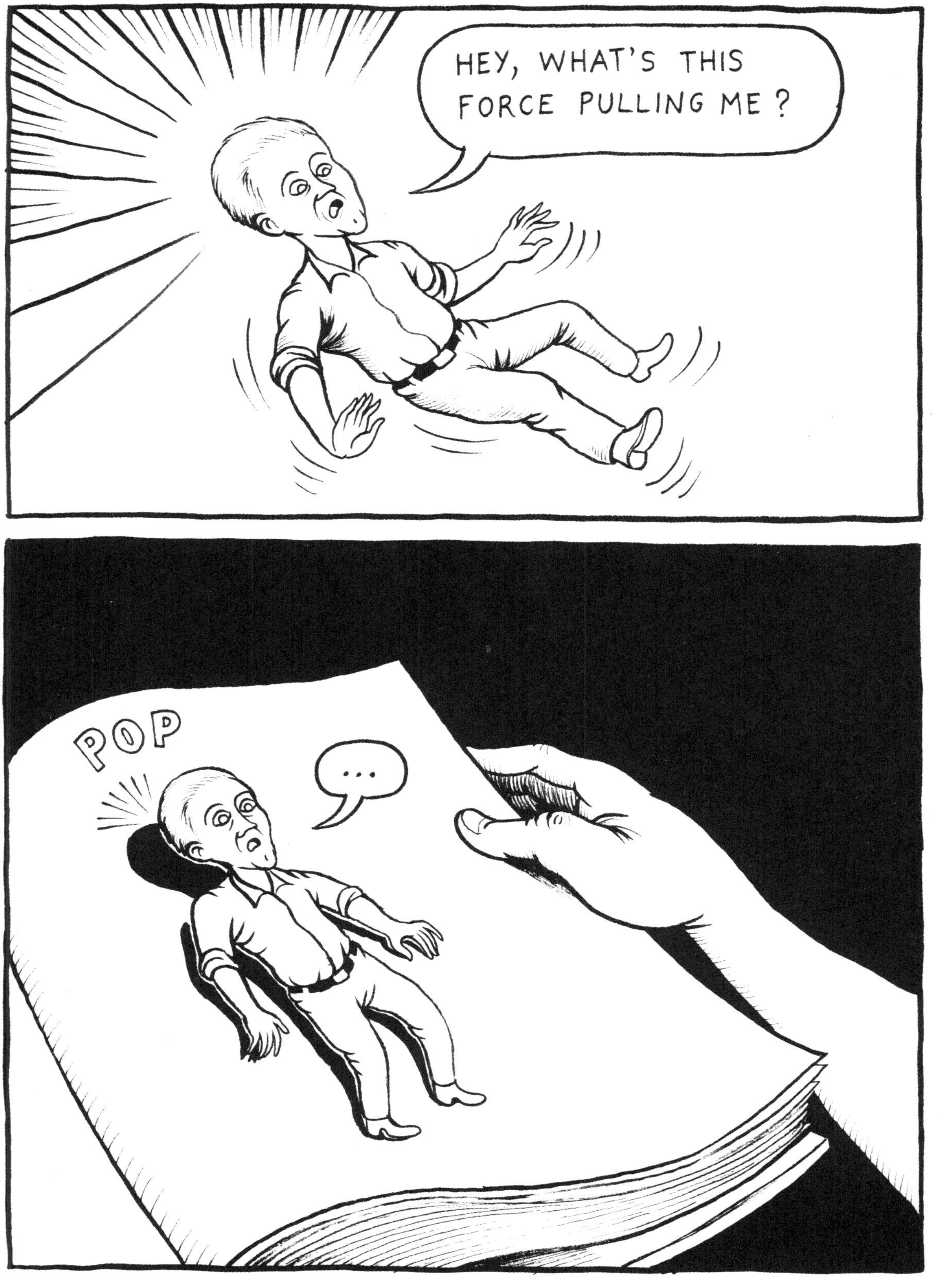
HEY, WHAT'S THIS FORCE PULLING ME?
POP
...

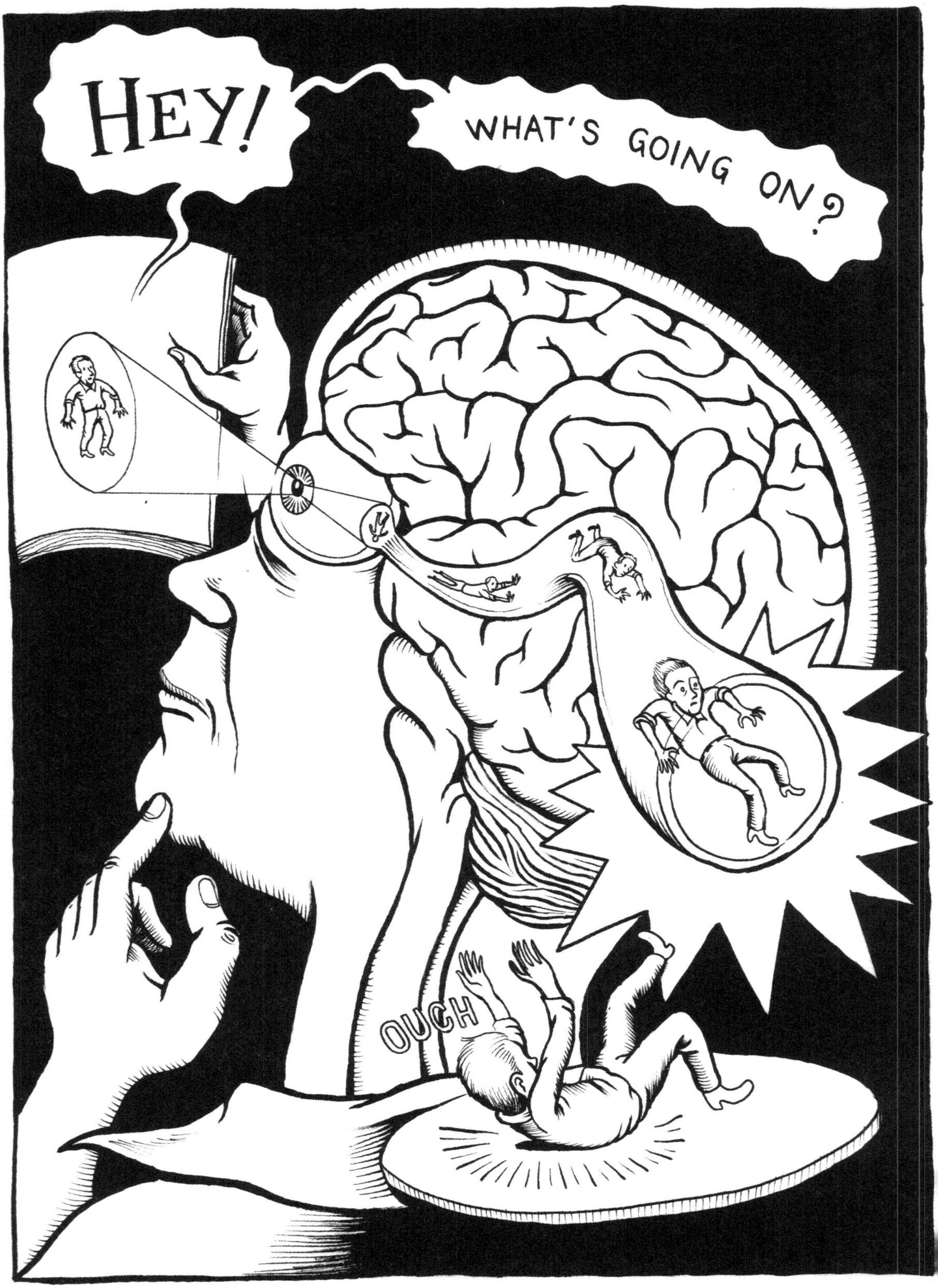
HEY!
WHAT'S GOING ON?
OUCH

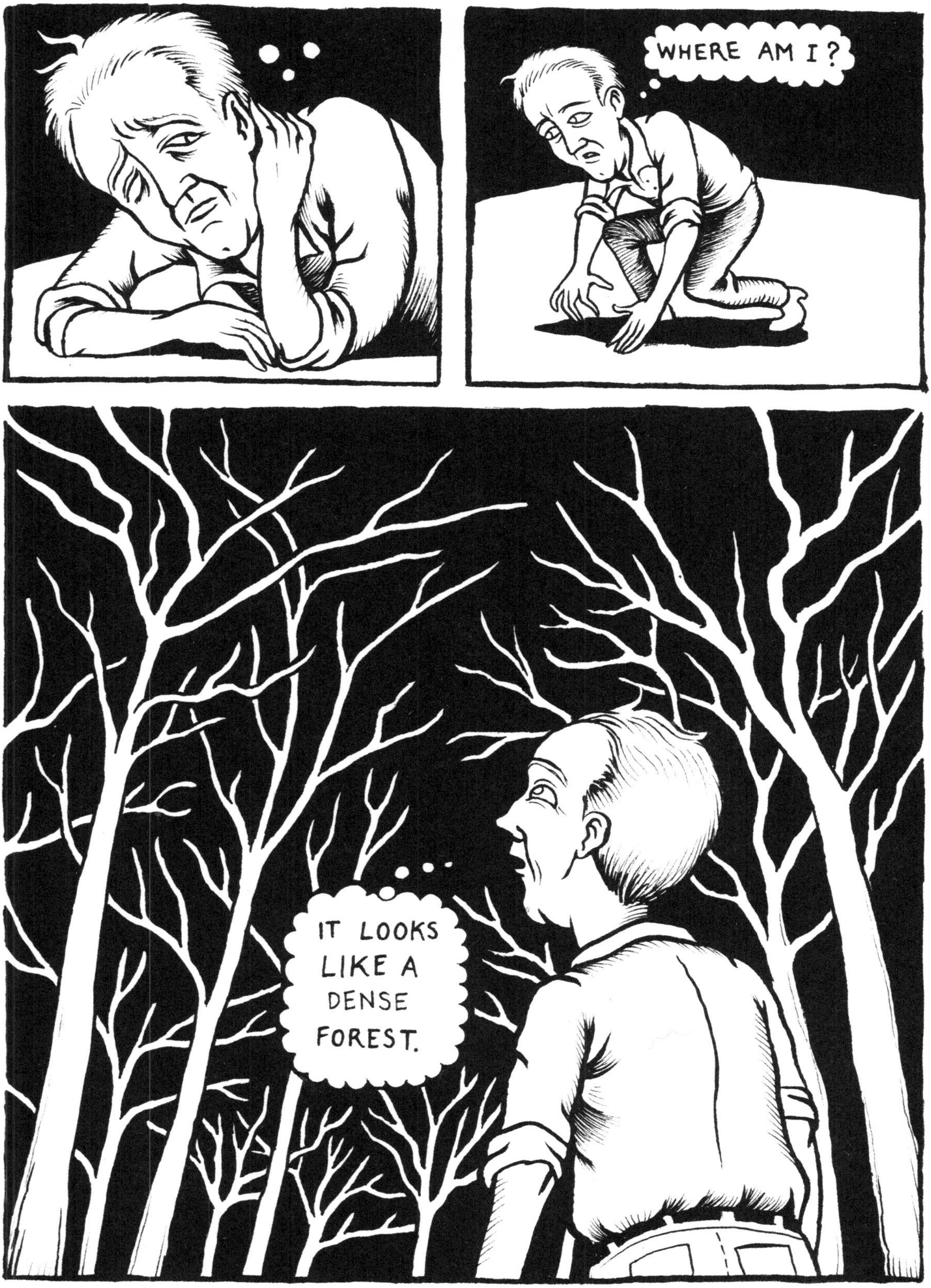
WHERE AM I?
IT LOOKS LIKE A DENSE FOREST.

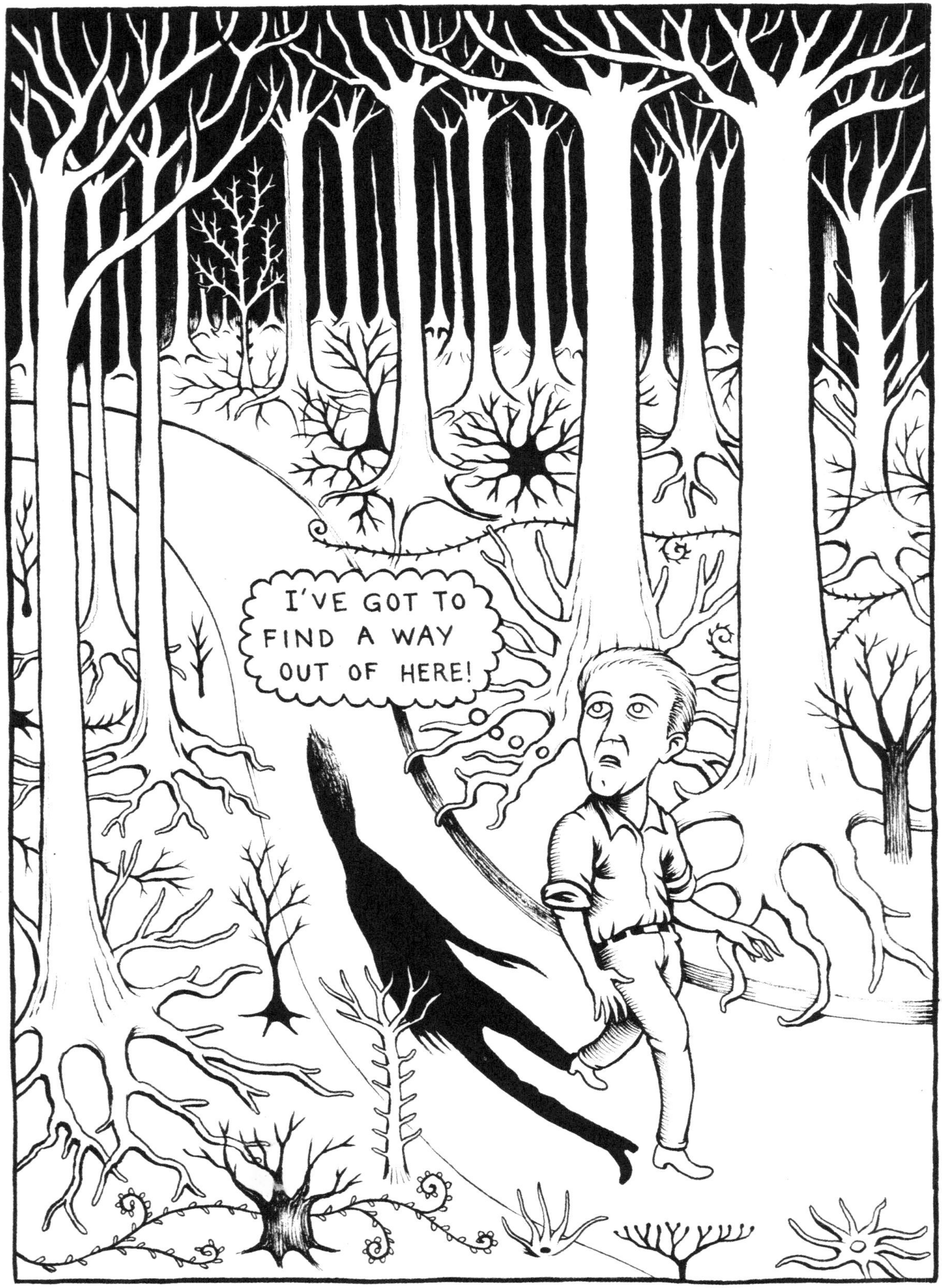
I'VE GOT TO FIND A WAY OUT OF HERE!

MORPHOLOGY

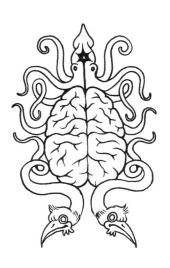

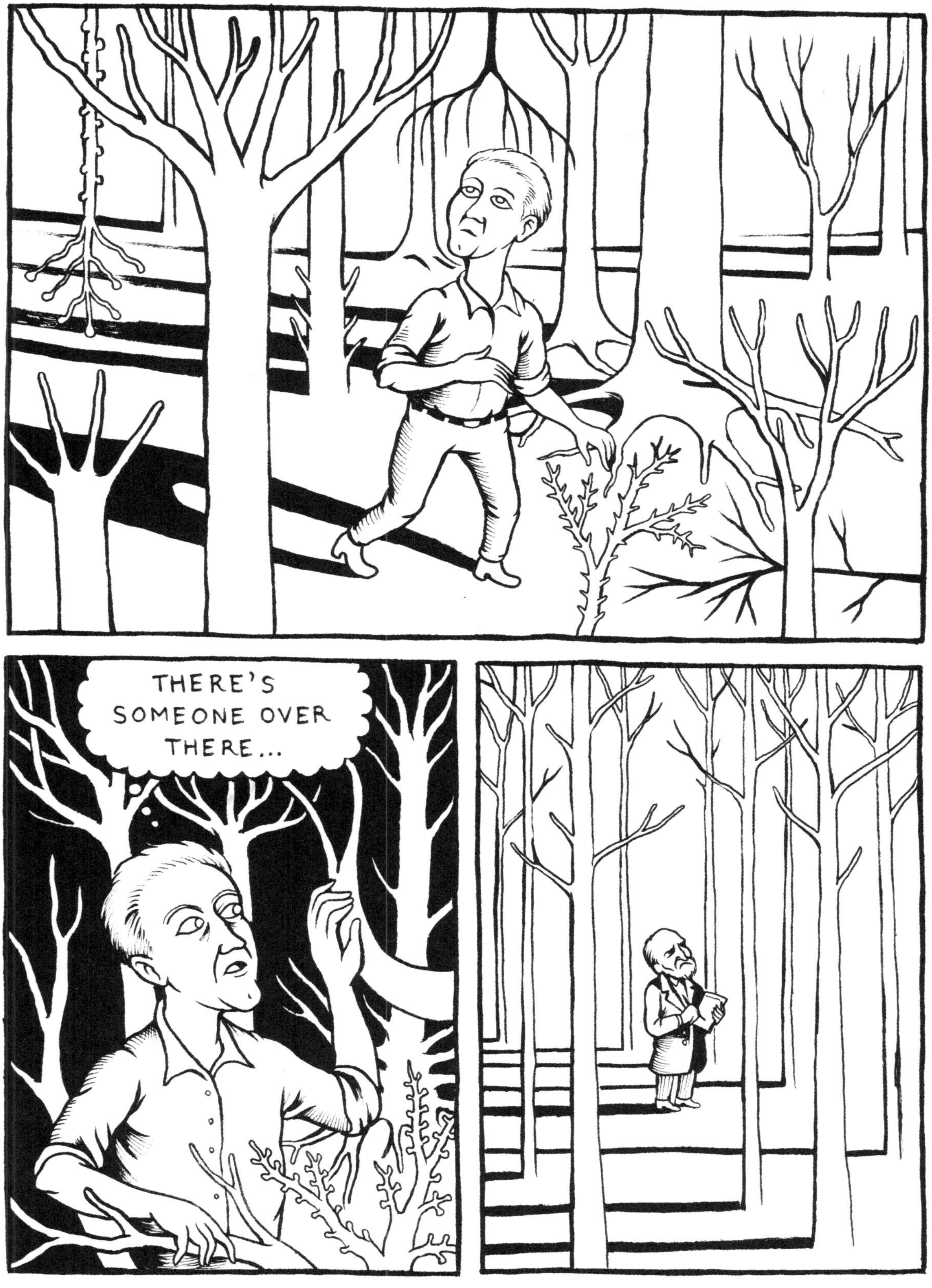
THERE'S SOMEONE OVER THERE...

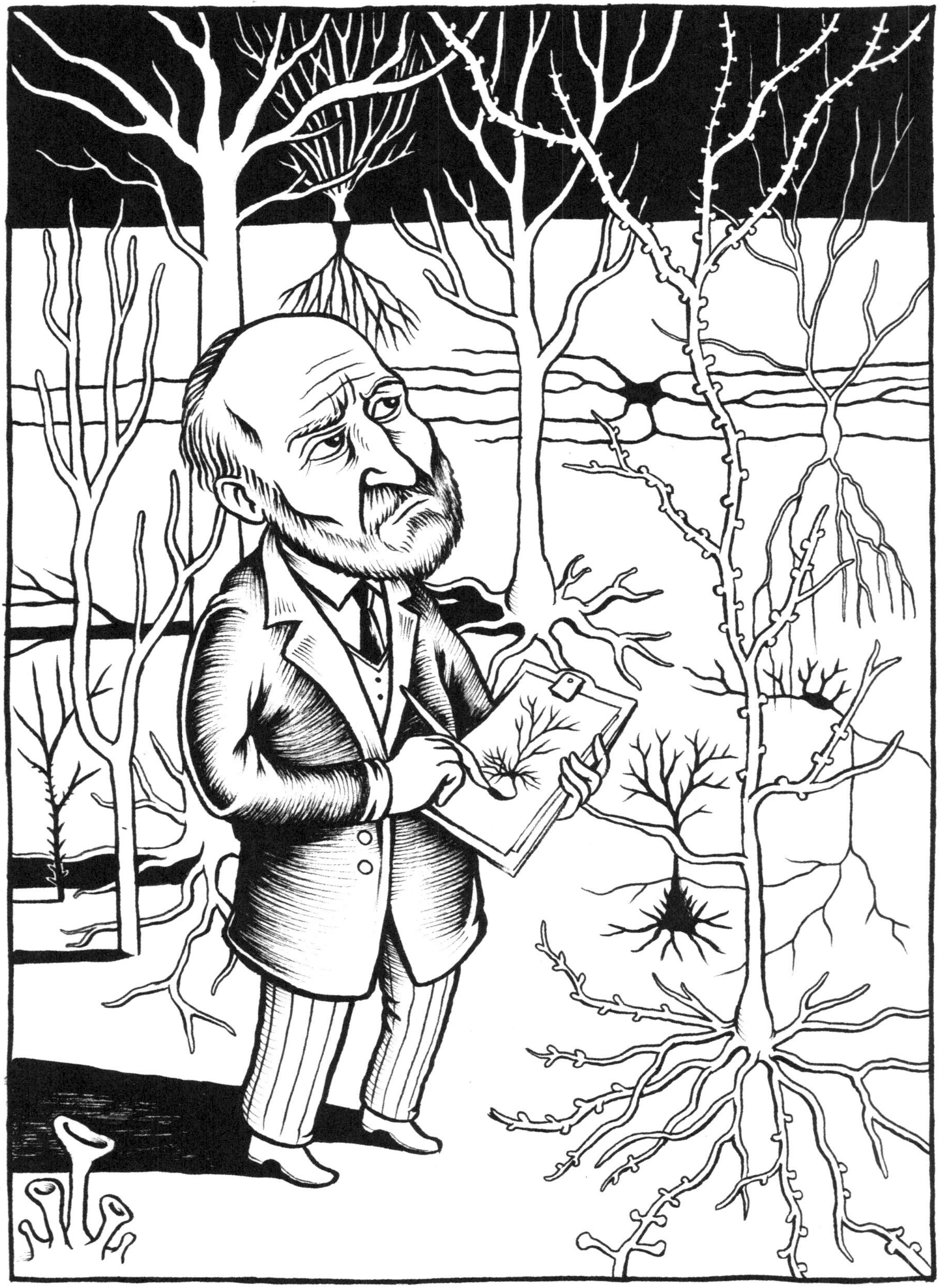

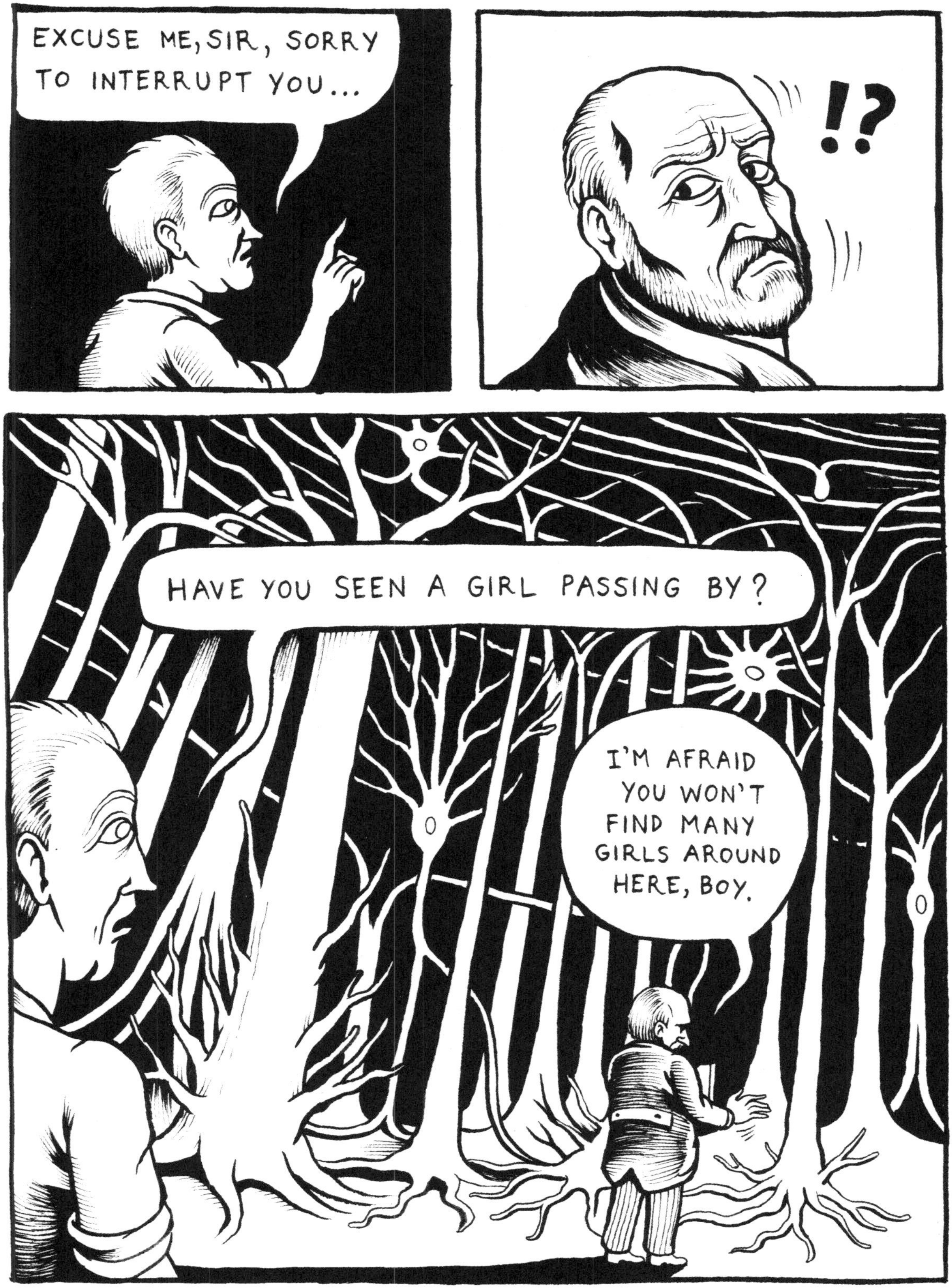
EXCUSE ME, SIR, SORRY TO INTERRUPT YOU...
!?
HAVE YOU SEEN A GIRL PASSING BY?
I'M AFRAID YOU WON'T FIND MANY GIRLS AROUND HERE, BOY.

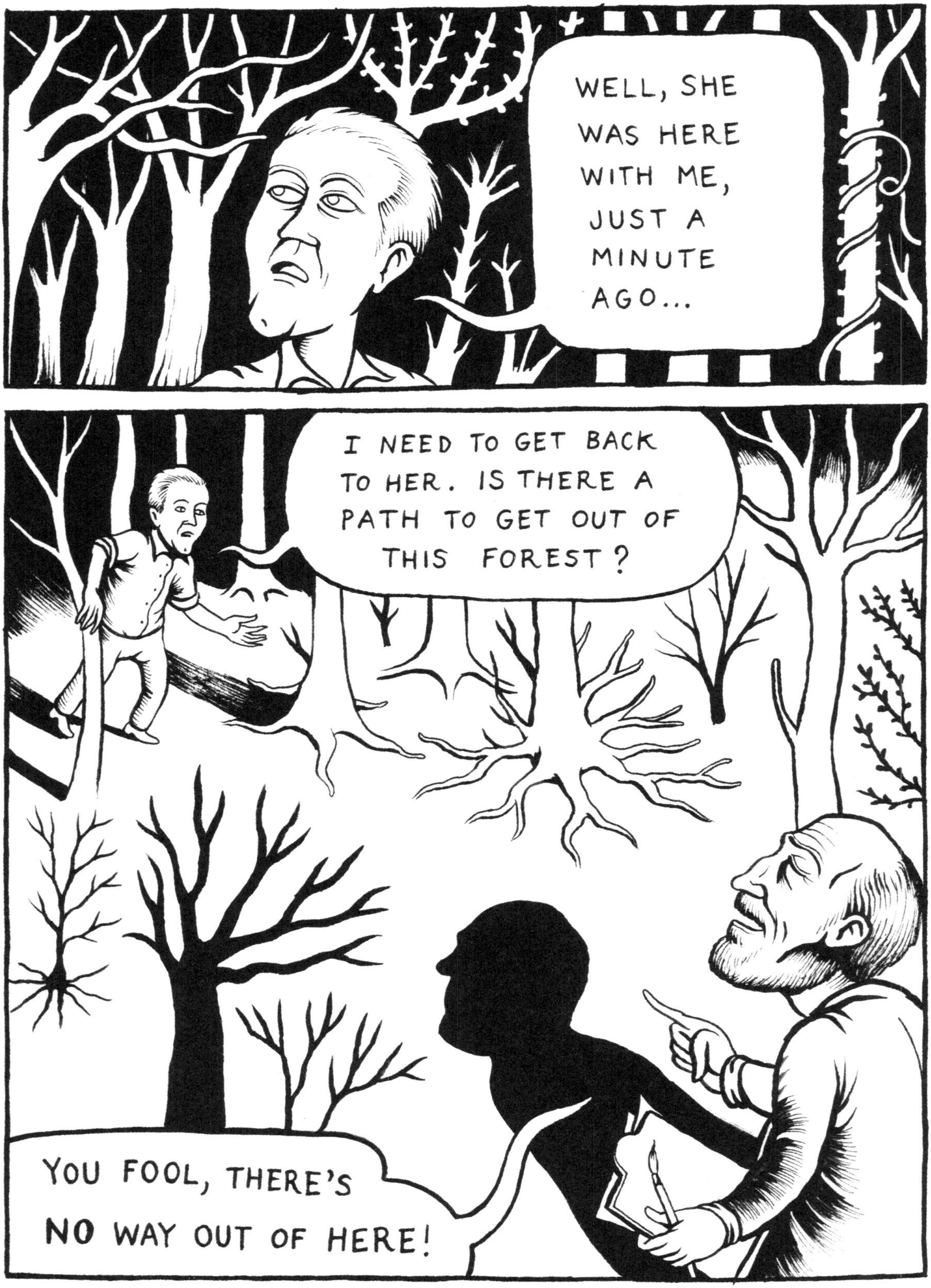
WELL, SHE WAS HERE WITH ME, JUST A MINUTE AGO...
I NEED TO GET BACK TO HER. IS THERE A PATH TO GET OUT OF THIS FOREST?
YOU FOOL, THERE'S **NO** WAY OUT OF HERE!

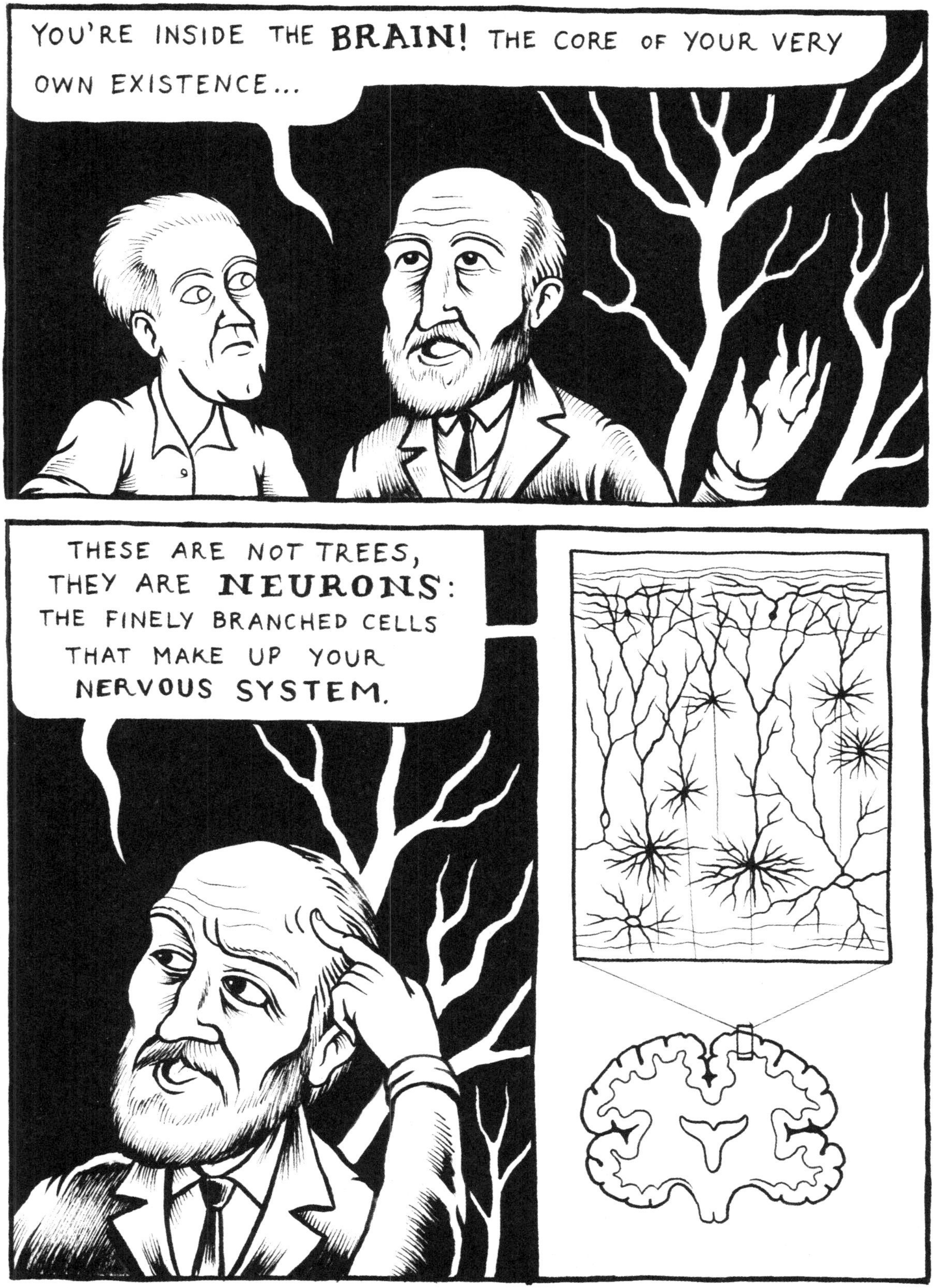
YOU'RE INSIDE THE BRAIN! THE CORE OF YOUR VERY OWN EXISTENCE...
THESE ARE NOT TREES, THEY ARE NEURONS: THE FINELY BRANCHED CELLS THAT MAKE UP YOUR NERVOUS SYSTEM.

IT ALL BEGINS AND ENDS WITH NEURONS: FROM YOUR SENSORY RECEPTORS TO THE NERVES THAT CONTROL YOUR MUSCLES. EVERYTHING YOU FEEL, REMEMBER OR DREAM IS WRITTEN IN THESE CELLS.

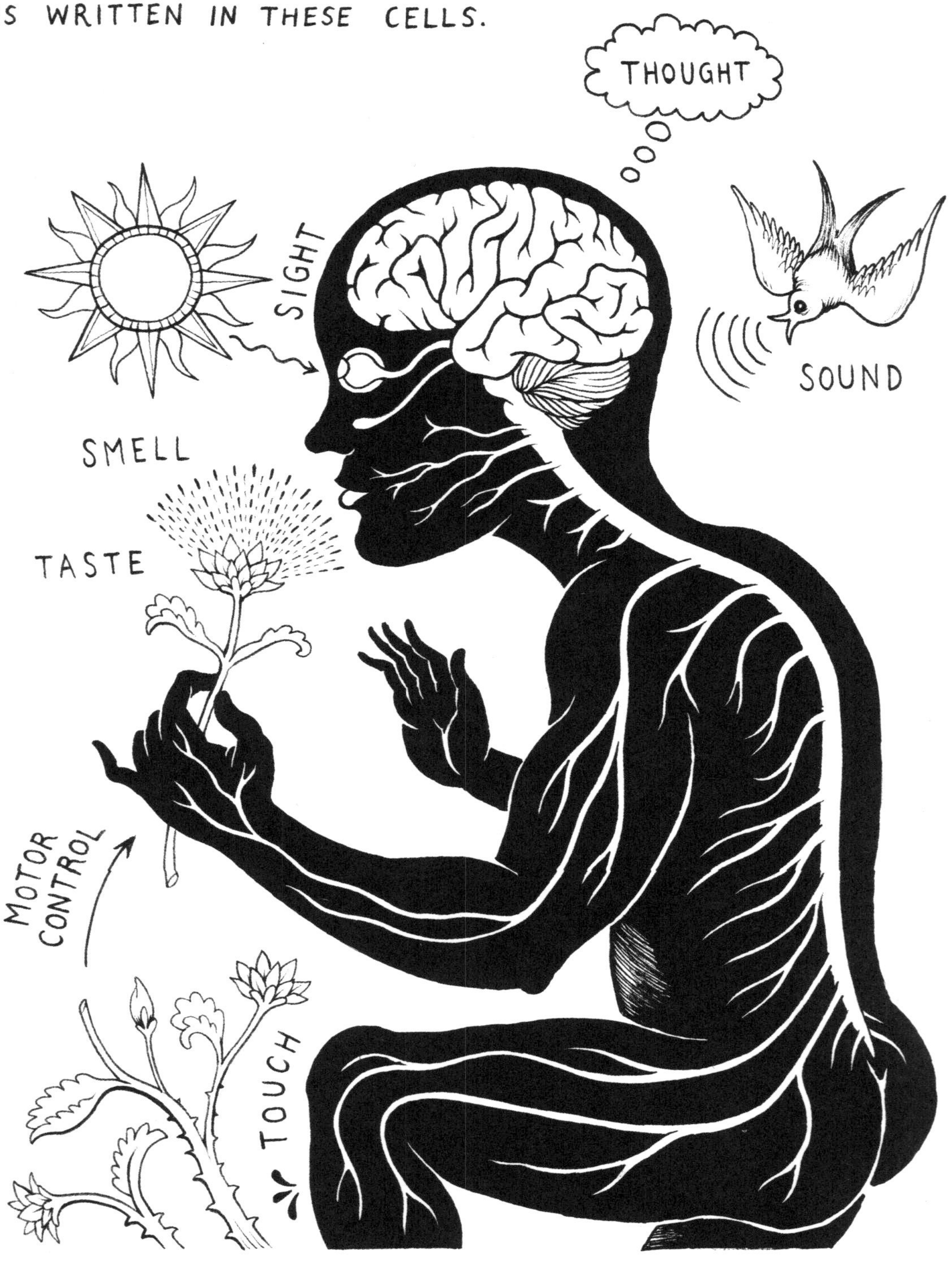
THOUGHT
SIGHT
SOUND
SMELL
TASTE
MOTOR CONTROL
TOUCH

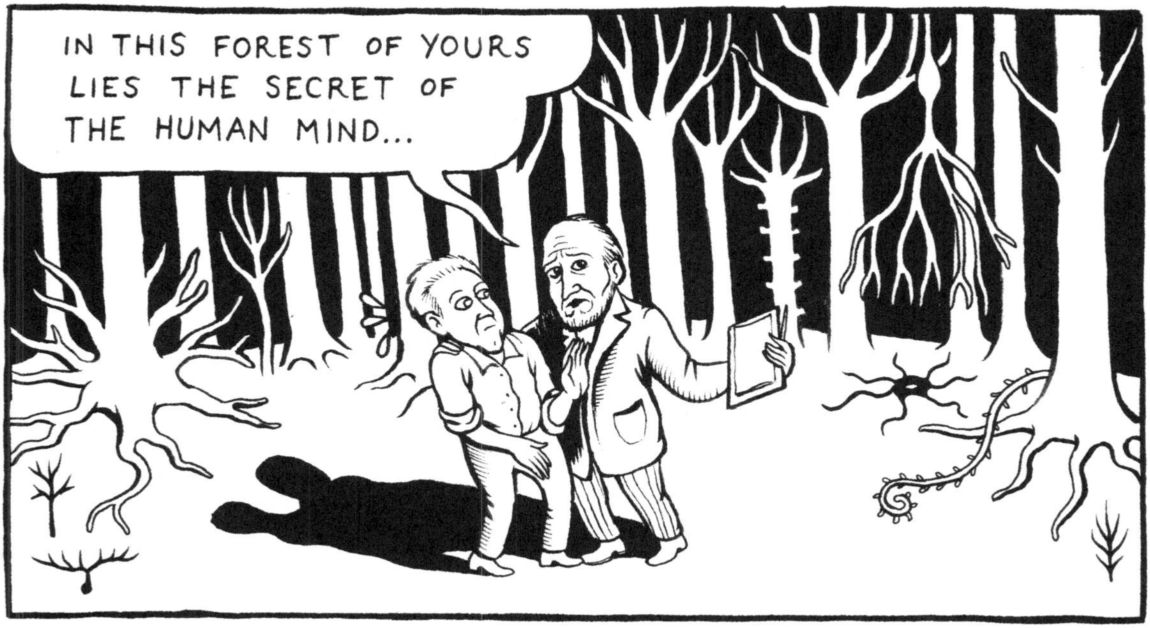
IN THIS FOREST OF YOURS LIES THE SECRET OF THE HUMAN MIND...

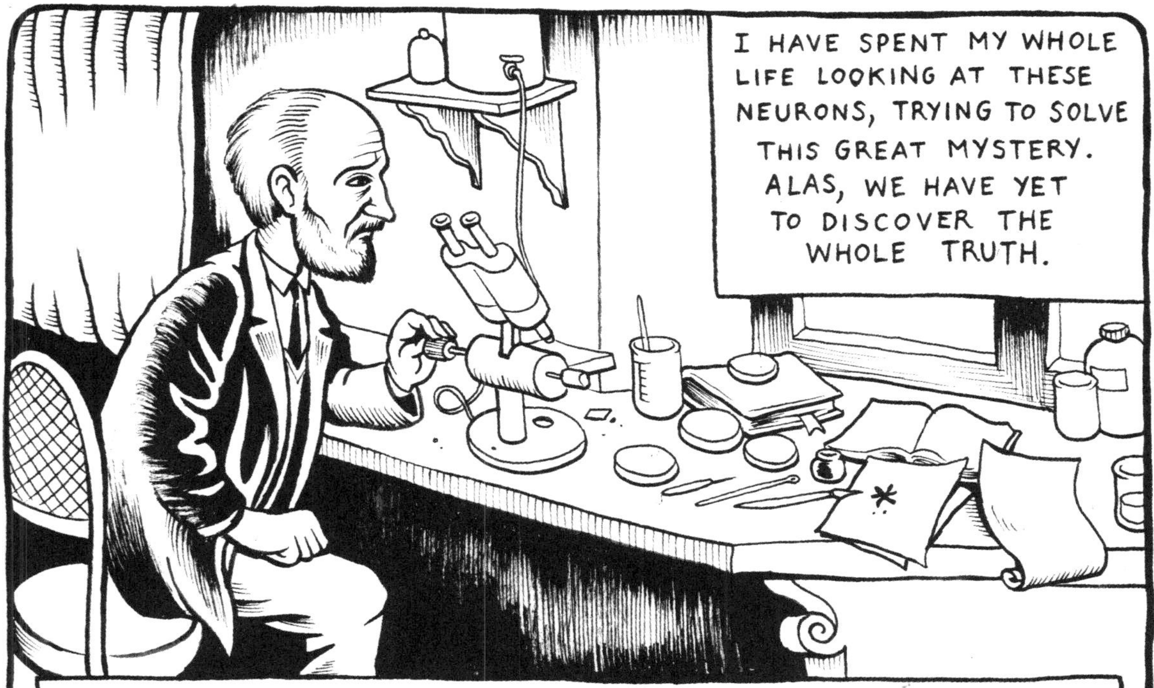
I HAVE SPENT MY WHOLE LIFE LOOKING AT THESE NEURONS, TRYING TO SOLVE THIS GREAT MYSTERY. ALAS, WE HAVE YET TO DISCOVER THE WHOLE TRUTH.

Santiago Ramón y Cajal (1852-1934) was a Spanish neuroscientist and Nobel laureate. His pioneering investigations of the structure of the brain have led him to be considered as the father of neuroscience, although he always had a great passion for drawing.

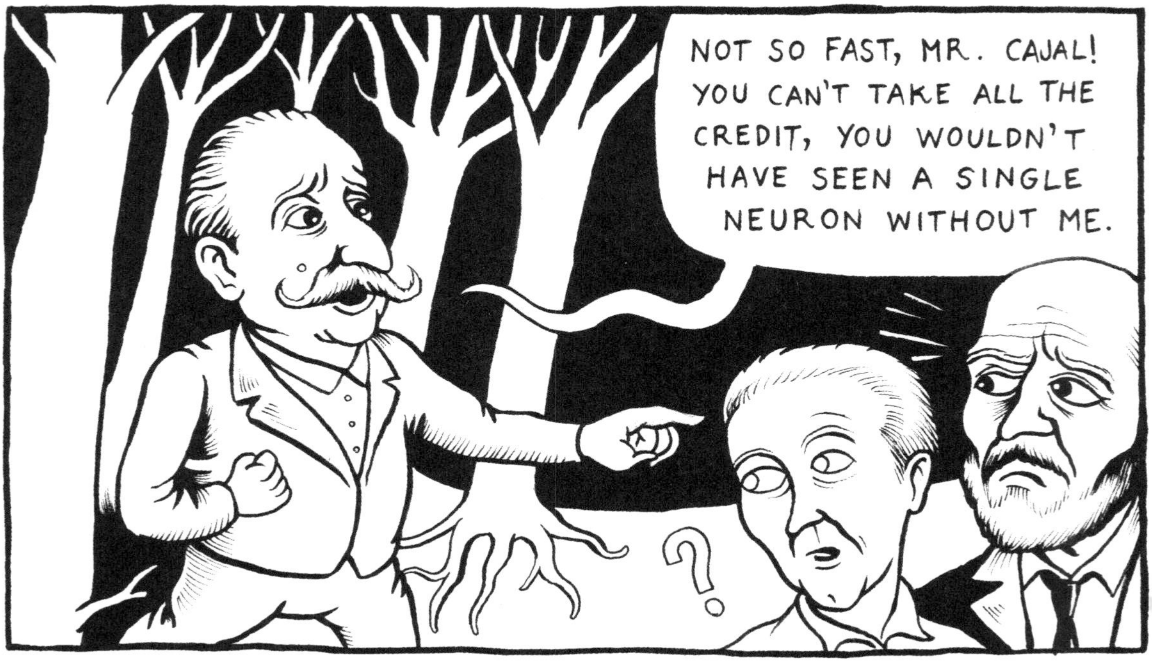
NOT SO FAST, MR. CAJAL! YOU CAN'T TAKE ALL THE CREDIT, YOU WOULDN'T HAVE SEEN A SINGLE NEURON WITHOUT ME.

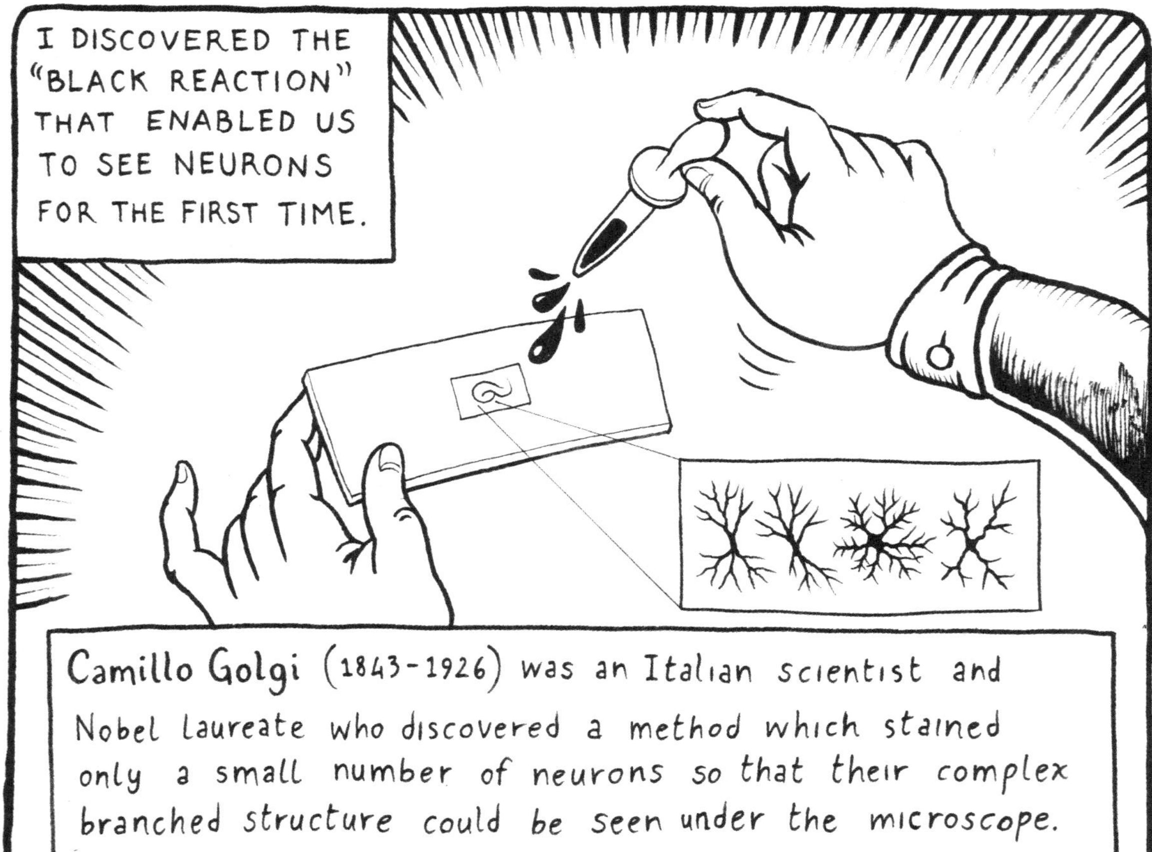
I DISCOVERED THE "BLACK REACTION" THAT ENABLED US TO SEE NEURONS FOR THE FIRST TIME.
Camillo Golgi (1843-1926) was an Italian scientist and Nobel laureate who discovered a method which stained only a small number of neurons so that their complex branched structure could be seen under the microscope.

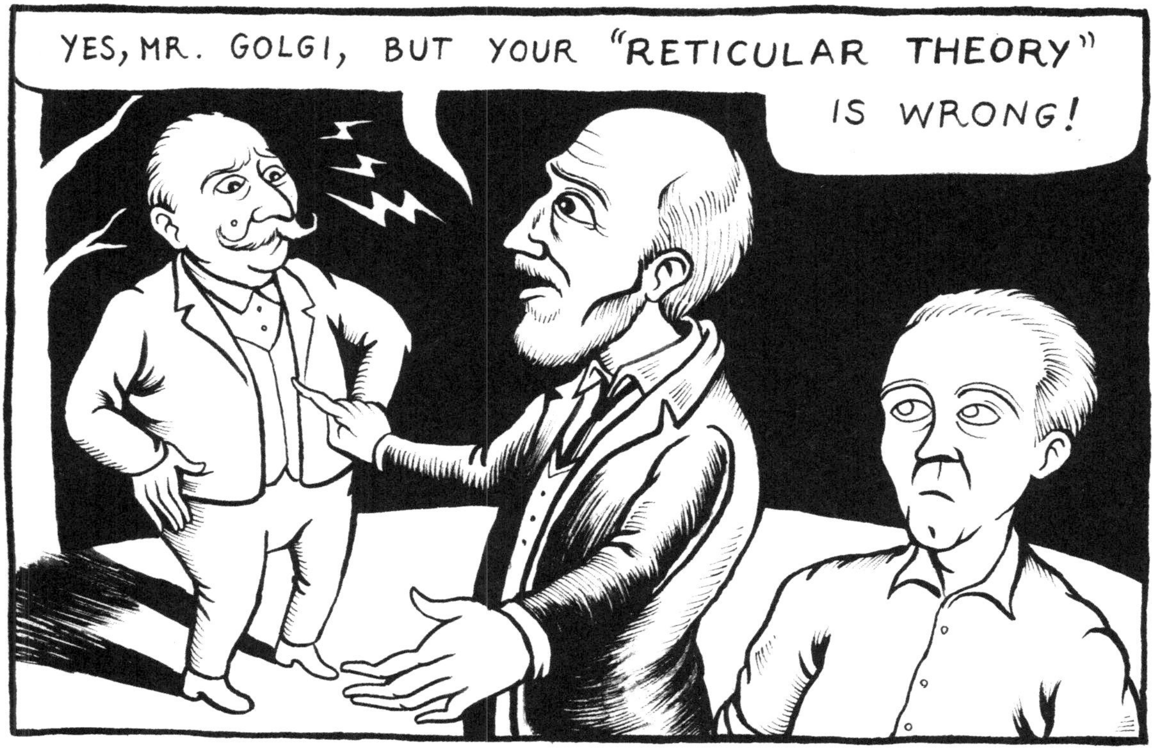
YES, MR. GOLGI, BUT YOUR "RETICULAR THEORY" IS WRONG!

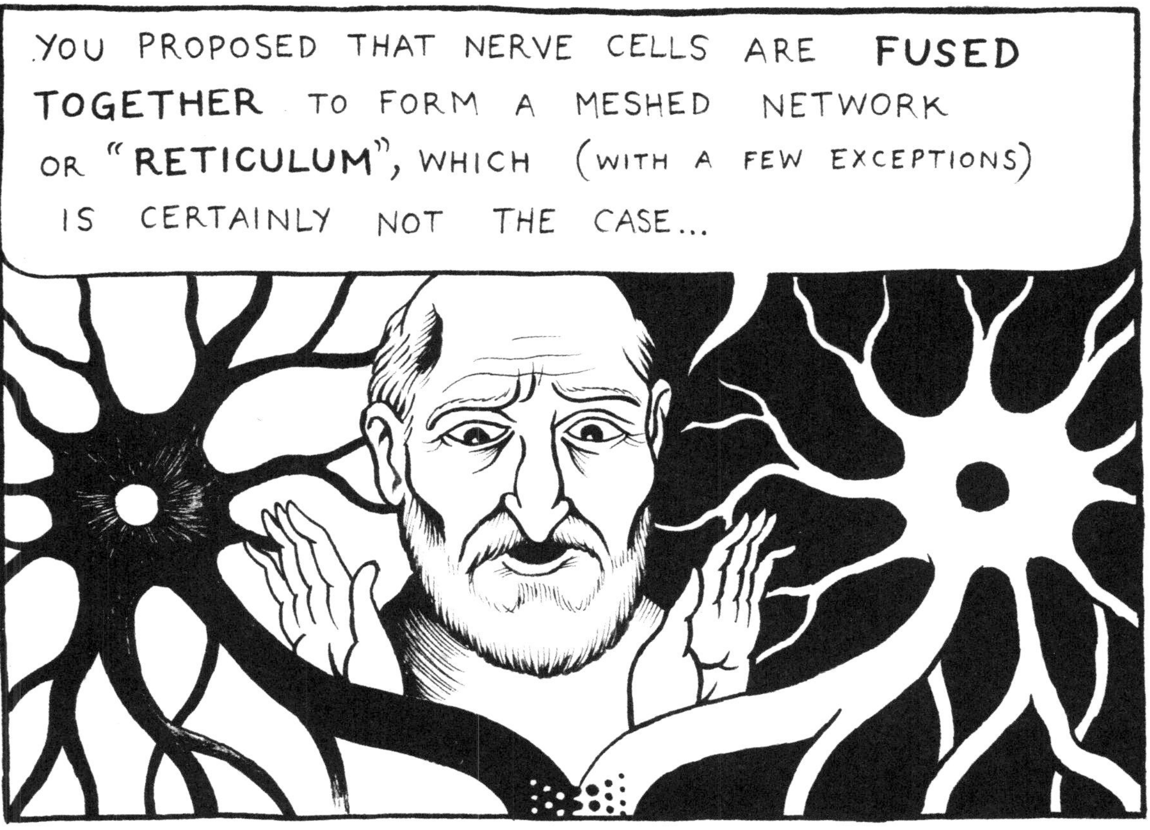
YOU PROPOSED THAT NERVE CELLS ARE FUSED TOGETHER TO FORM A MESHED NETWORK OR "RETICULUM", WHICH (WITH A FEW EXCEPTIONS) IS CERTAINLY NOT THE CASE...

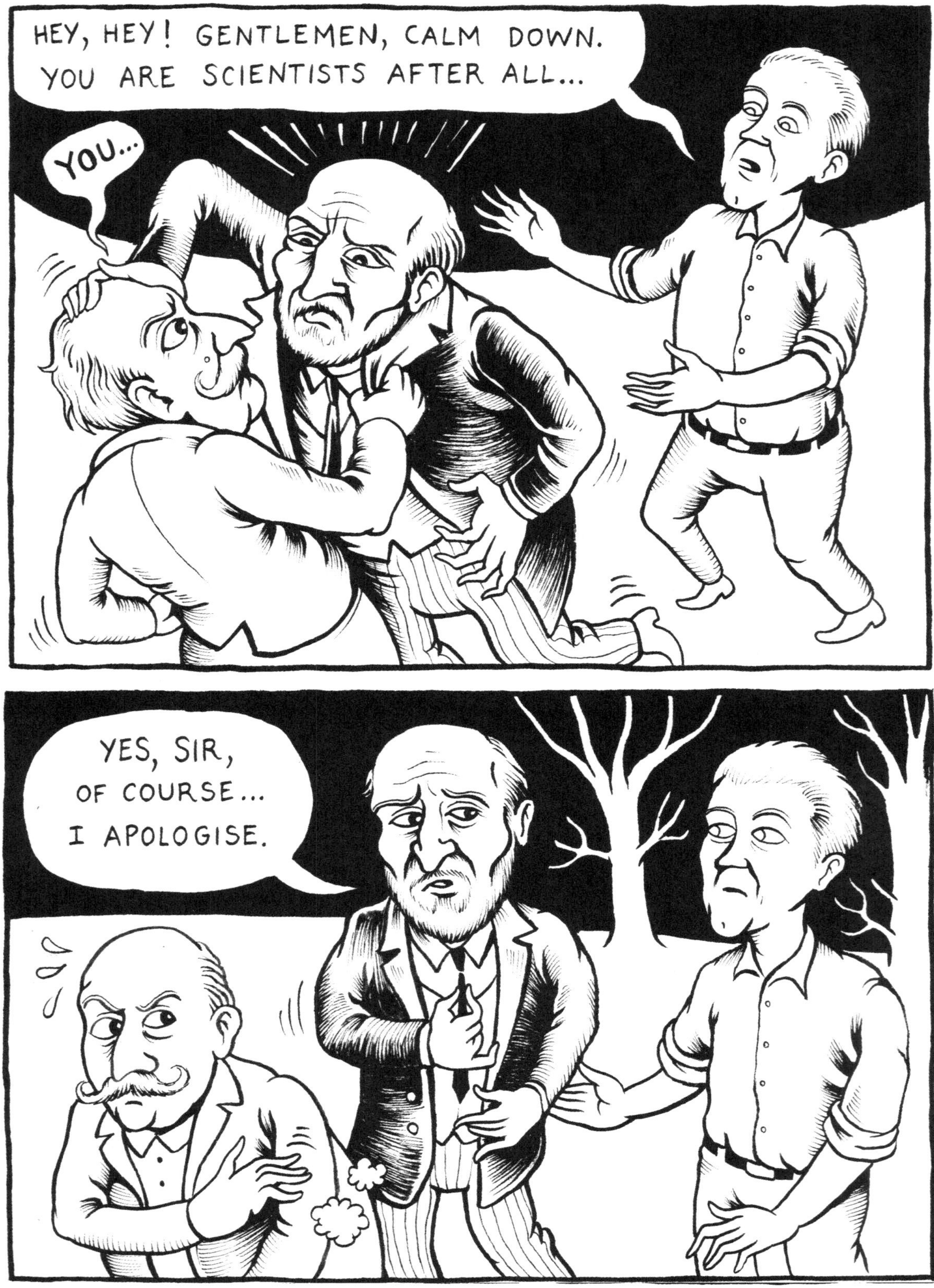
HEY, HEY! GENTLEMEN, CALM DOWN. YOU ARE SCIENTISTS AFTER ALL...
YOU...
YES, SIR, OF COURSE... I APOLOGISE.

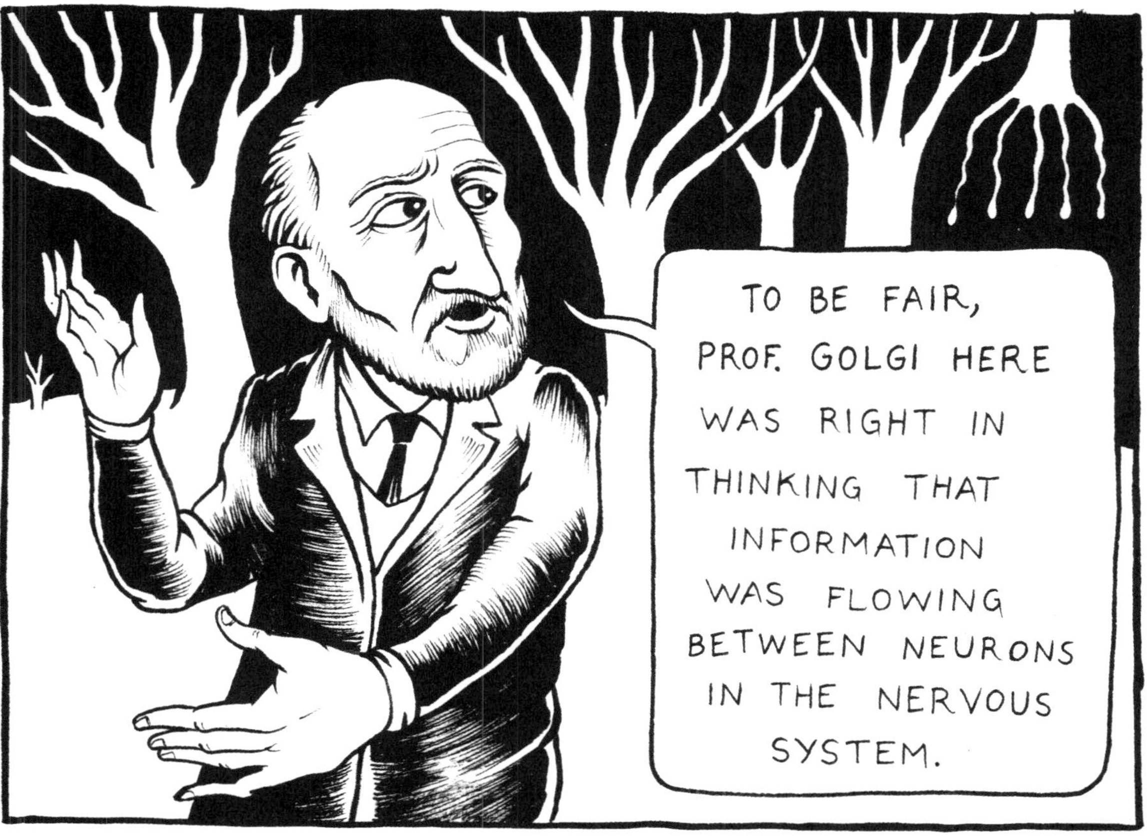
TO BE FAIR, PROF. GOLGI HERE WAS RIGHT IN THINKING THAT INFORMATION WAS FLOWING BETWEEN NEURONS IN THE NERVOUS SYSTEM.

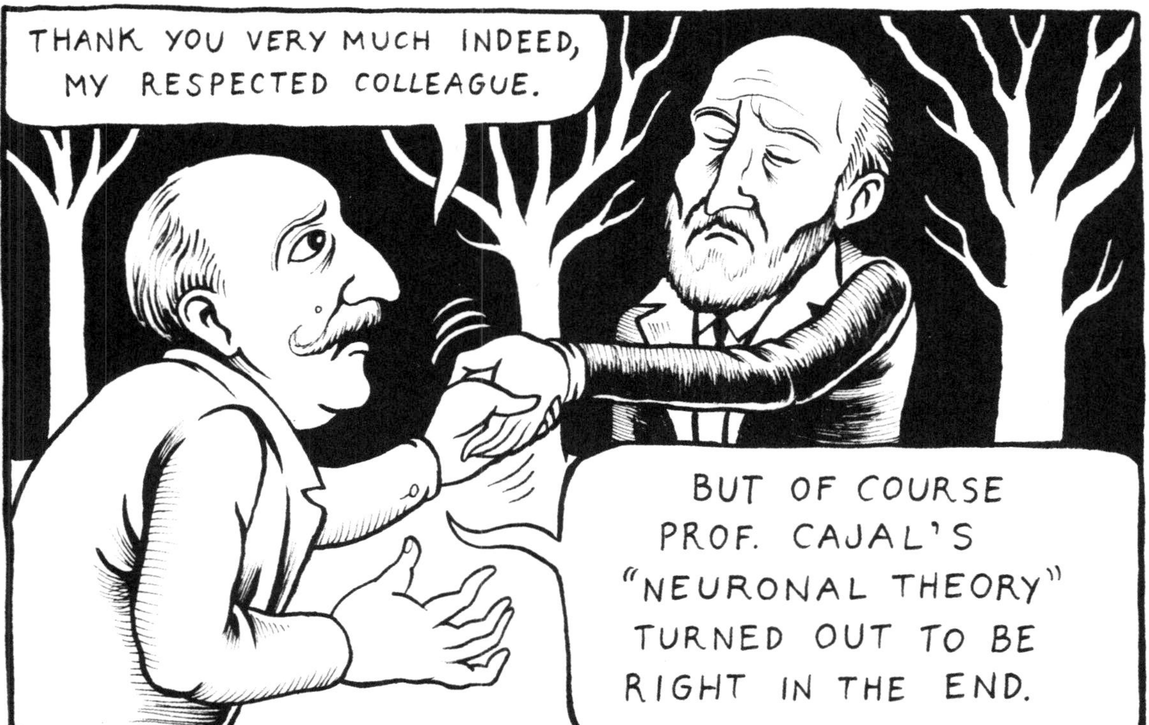
THANK YOU VERY MUCH INDEED, MY RESPECTED COLLEAGUE.
BUT OF COURSE PROF. CAJAL'S "NEURONAL THEORY" TURNED OUT TO BE RIGHT IN THE END.

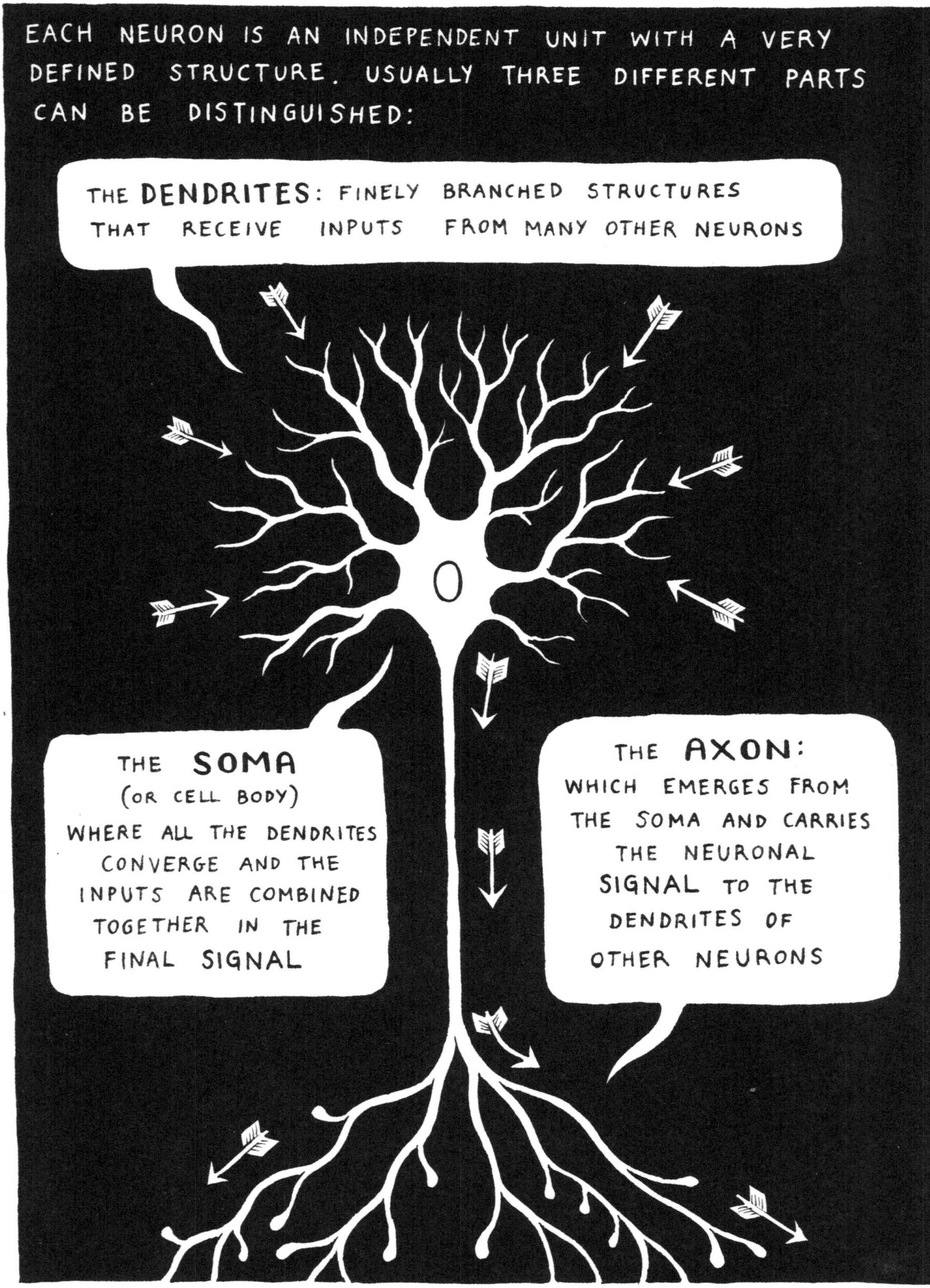
EACH NEURON IS AN INDEPENDENT UNIT WITH A VERY DEFINED STRUCTURE. USUALLY THREE DIFFERENT PARTS CAN BE DISTINGUISHED:
THE DENDRITES: FINELY BRANCHED STRUCTURES THAT RECEIVE INPUTS FROM MANY OTHER NEURONS
THE SOMA (OR CELL BODY) WHERE ALL THE DENDRITES CONVERGE AND THE INPUTS ARE COMBINED TOGETHER IN THE FINAL SIGNAL
THE AXON: WHICH EMERGES FROM THE SOMA AND CARRIES THE NEURONAL SIGNAL TO THE DENDRITES OF OTHER NEURONS

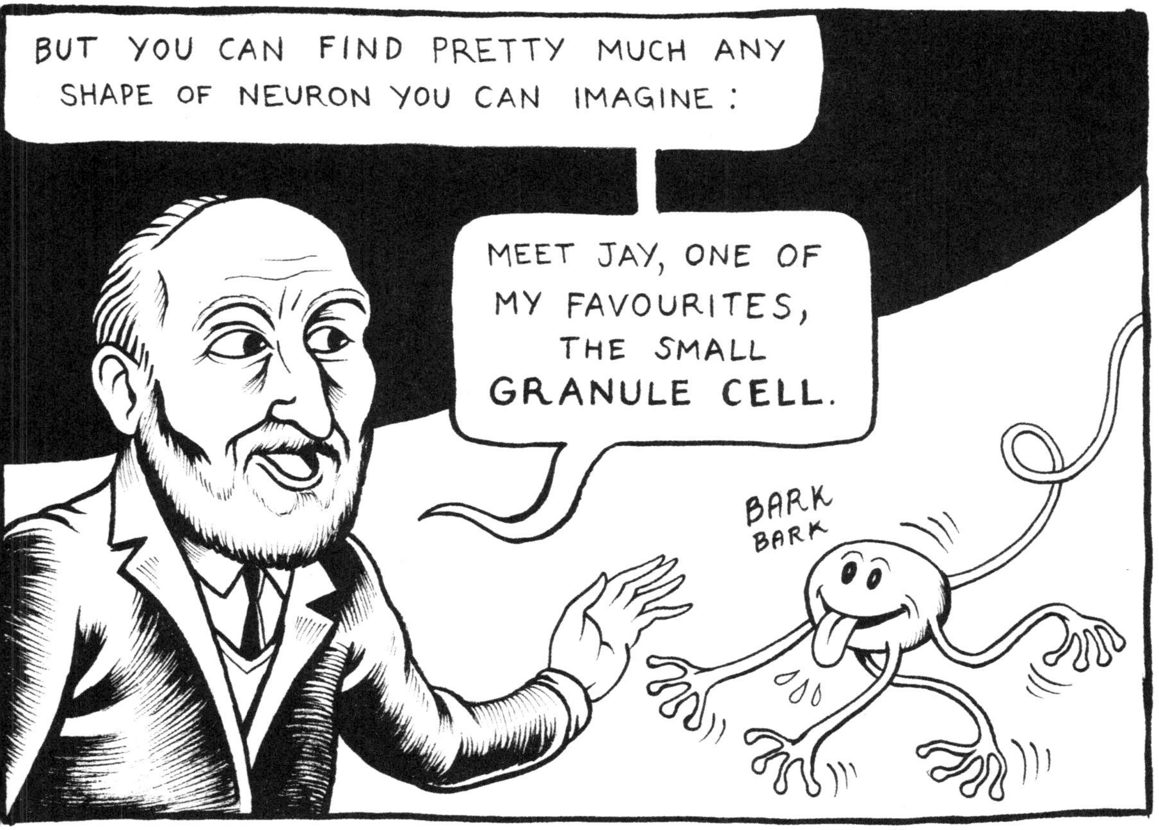
BUT YOU CAN FIND PRETTY MUCH ANY SHAPE OF NEURON YOU CAN IMAGINE:
MEET JAY, ONE OF MY FAVOURITES, THE SMALL **GRANULE CELL**.
BARK BARK

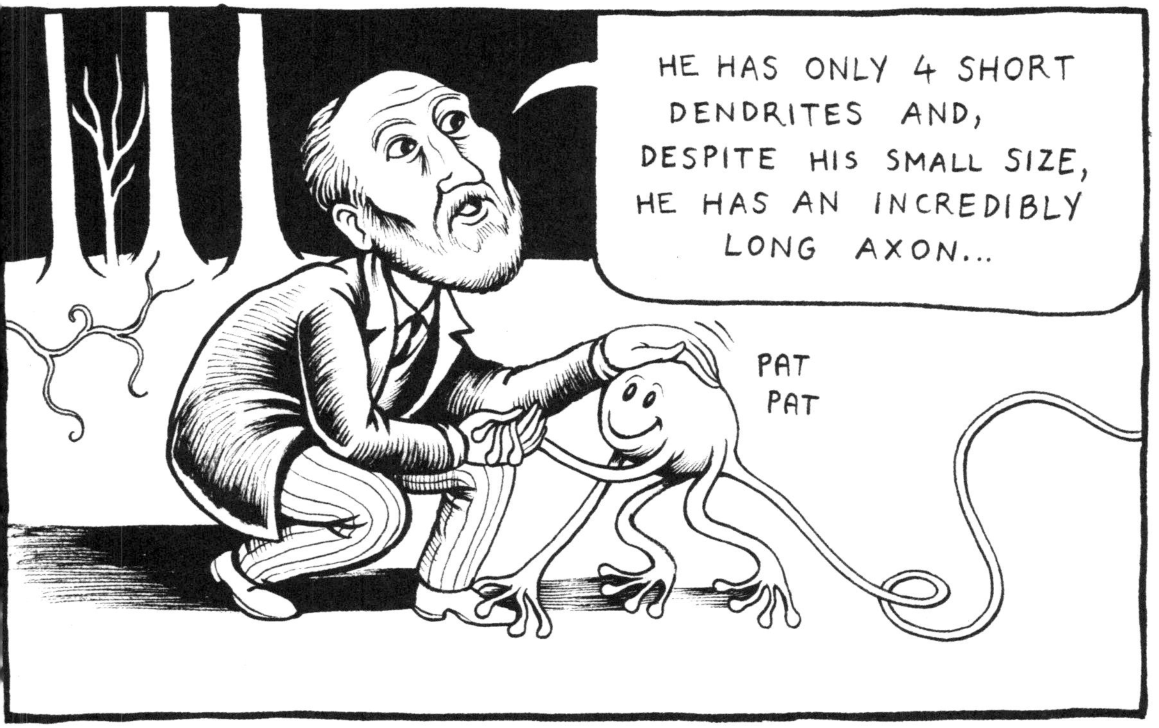
HE HAS ONLY 4 SHORT DENDRITES AND, DESPITE HIS SMALL SIZE, HE HAS AN INCREDIBLY LONG AXON...
PAT PAT

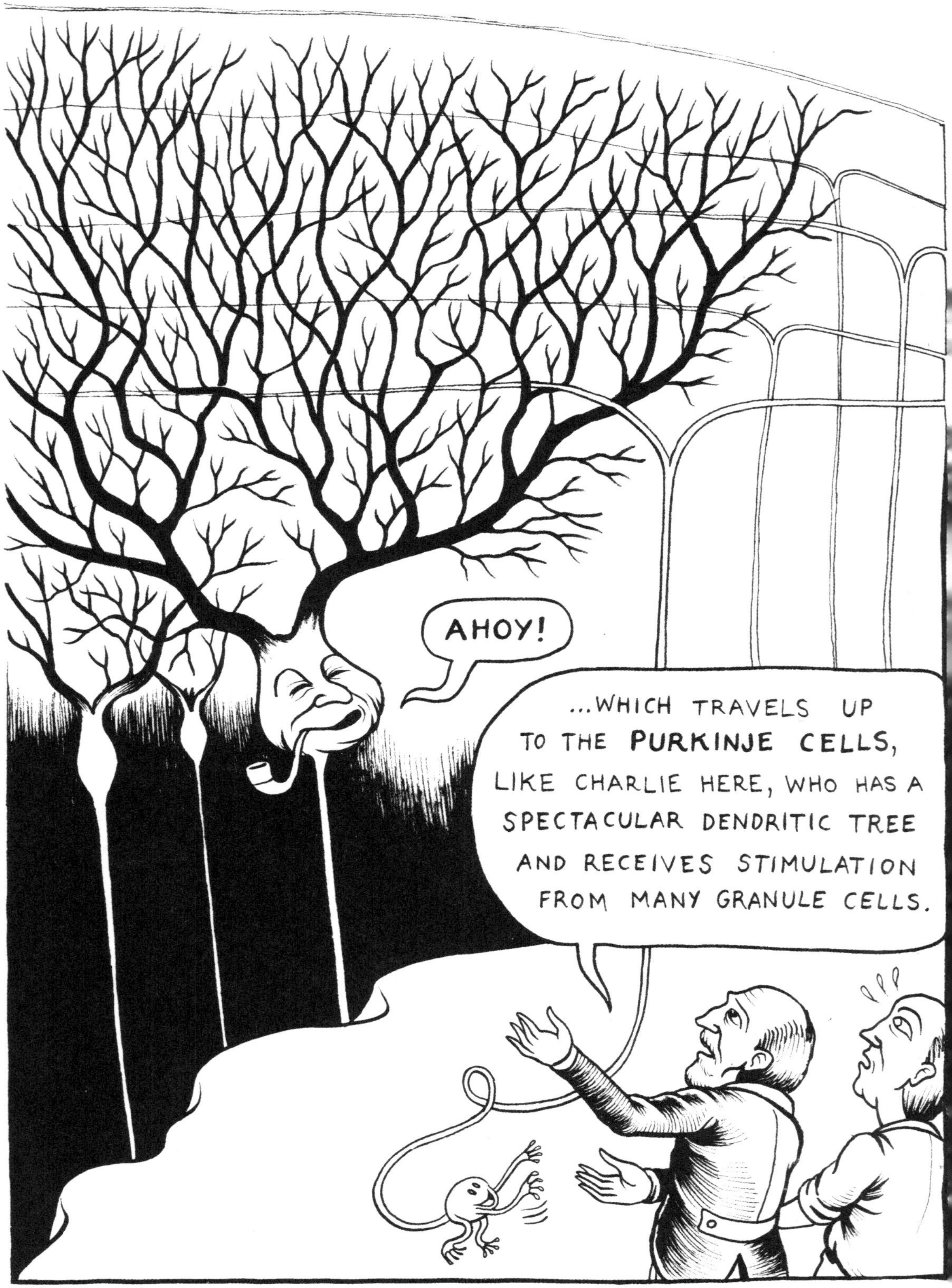
AHOY!
...WHICH TRAVELS UP TO THE PURKINJE CELLS, LIKE CHARLIE HERE, WHO HAS A SPECTACULAR DENDRITIC TREE AND RECEIVES STIMULATION FROM MANY GRANULE CELLS.

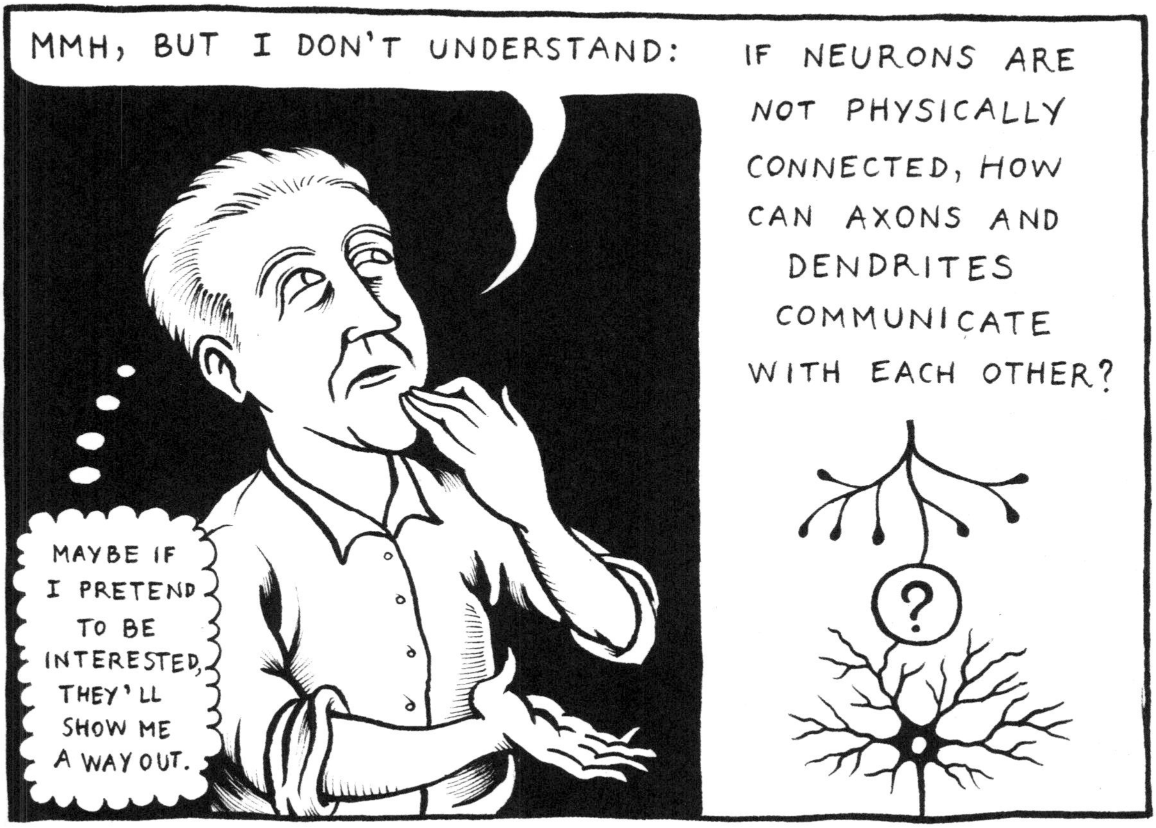
MMH, BUT I DON'T UNDERSTAND: IF NEURONS ARE NOT PHYSICALLY CONNECTED, HOW CAN AXONS AND DENDRITES COMMUNICATE WITH EACH OTHER?
MAYBE IF I PRETEND TO BE INTERESTED, THEY'LL SHOW ME A WAY OUT.
?

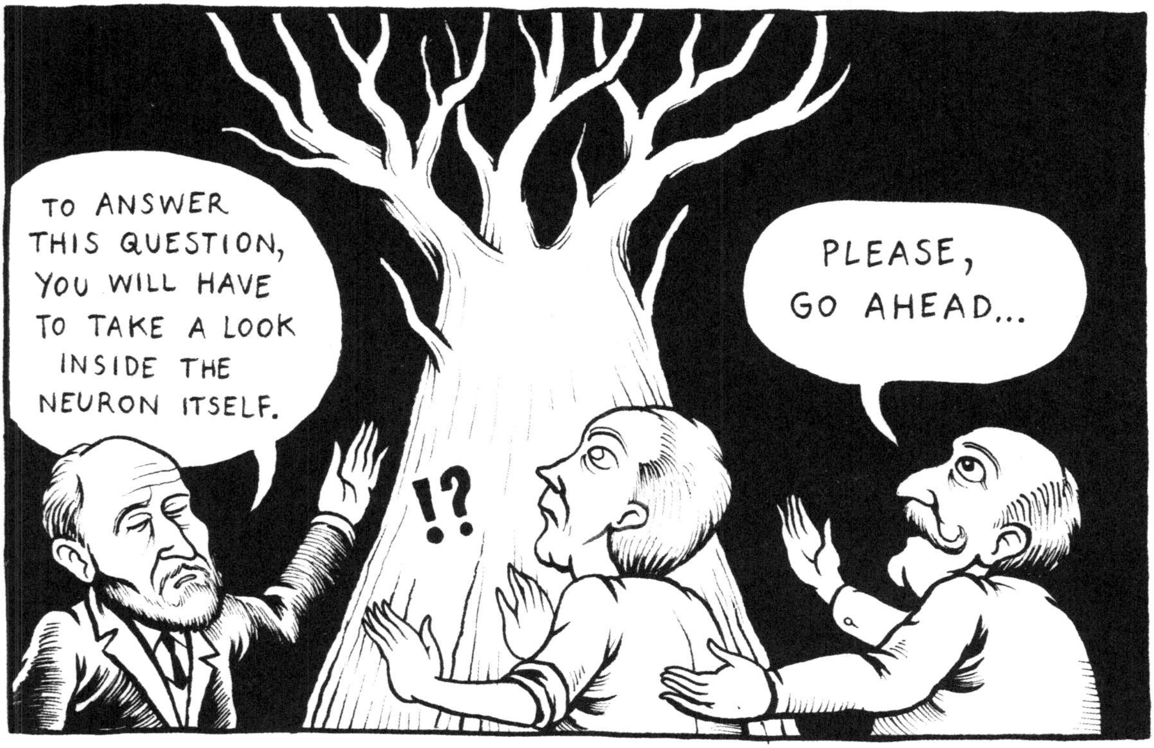
TO ANSWER THIS QUESTION, YOU WILL HAVE TO TAKE A LOOK INSIDE THE NEURON ITSELF.
!?
PLEASE, GO AHEAD...

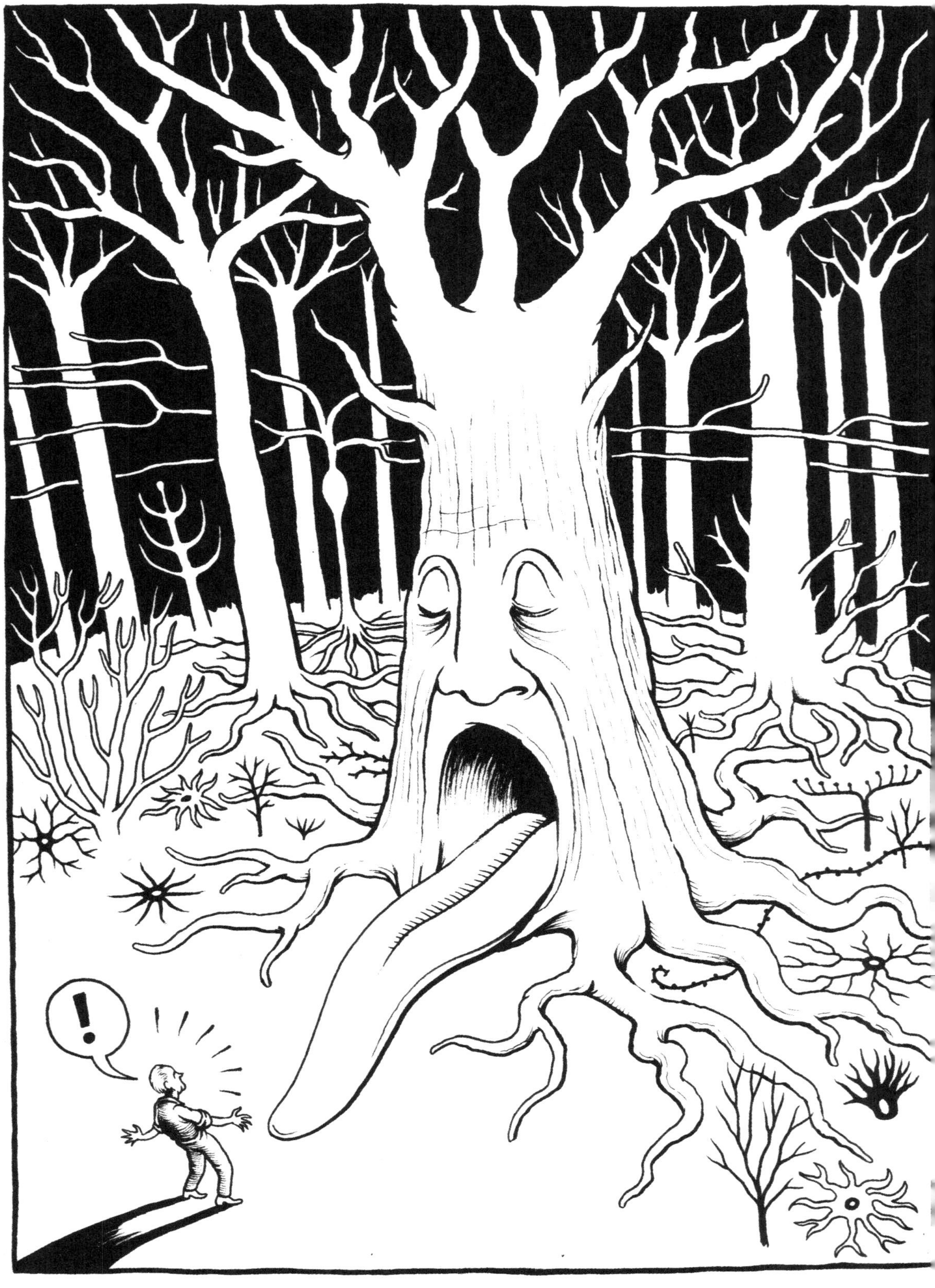
!

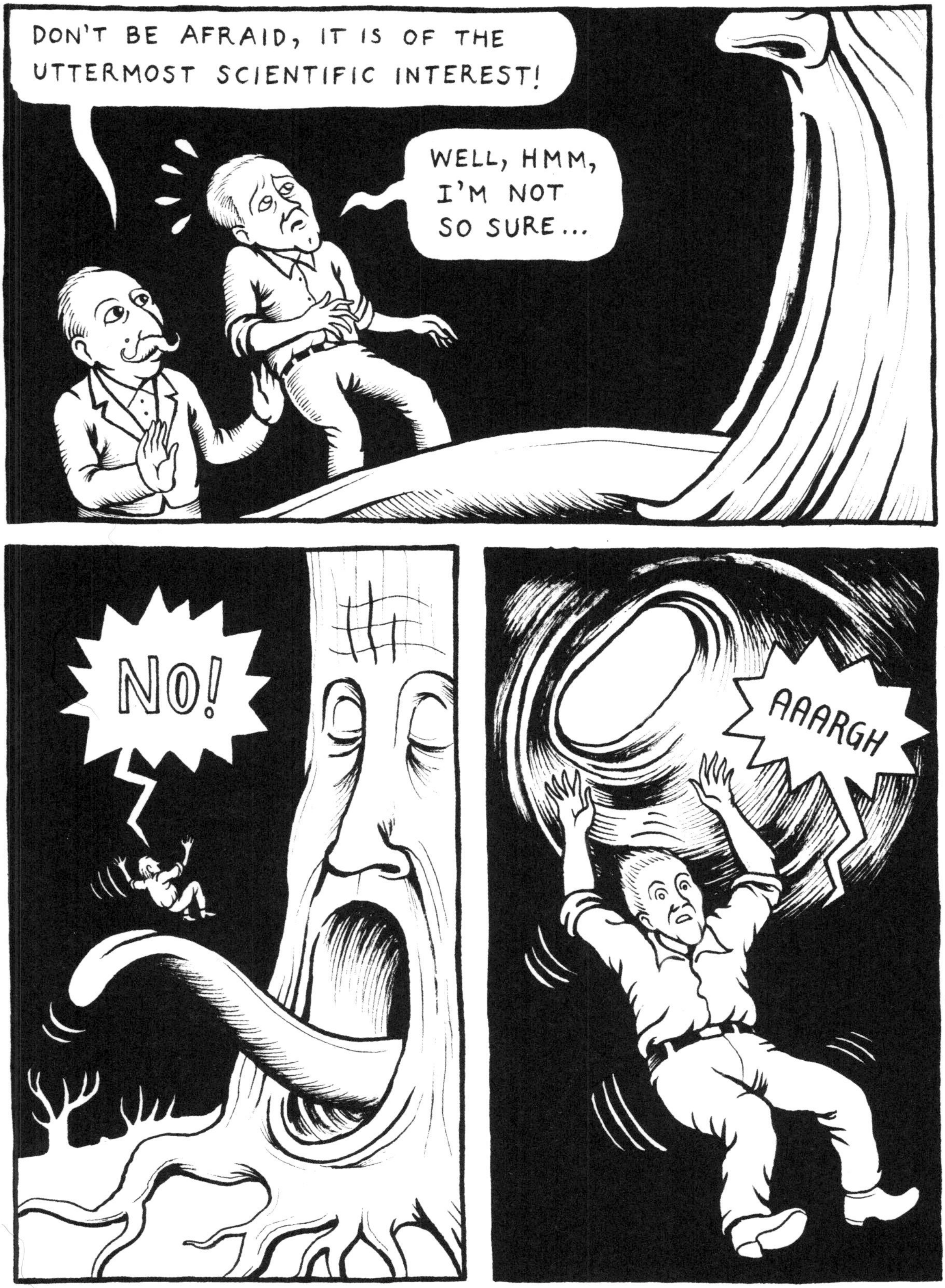
DON'T BE AFRAID, IT IS OF THE UTTERMOST SCIENTIFIC INTEREST!
WELL, HMM, I'M NOT SO SURE...
NO!
AAARGH

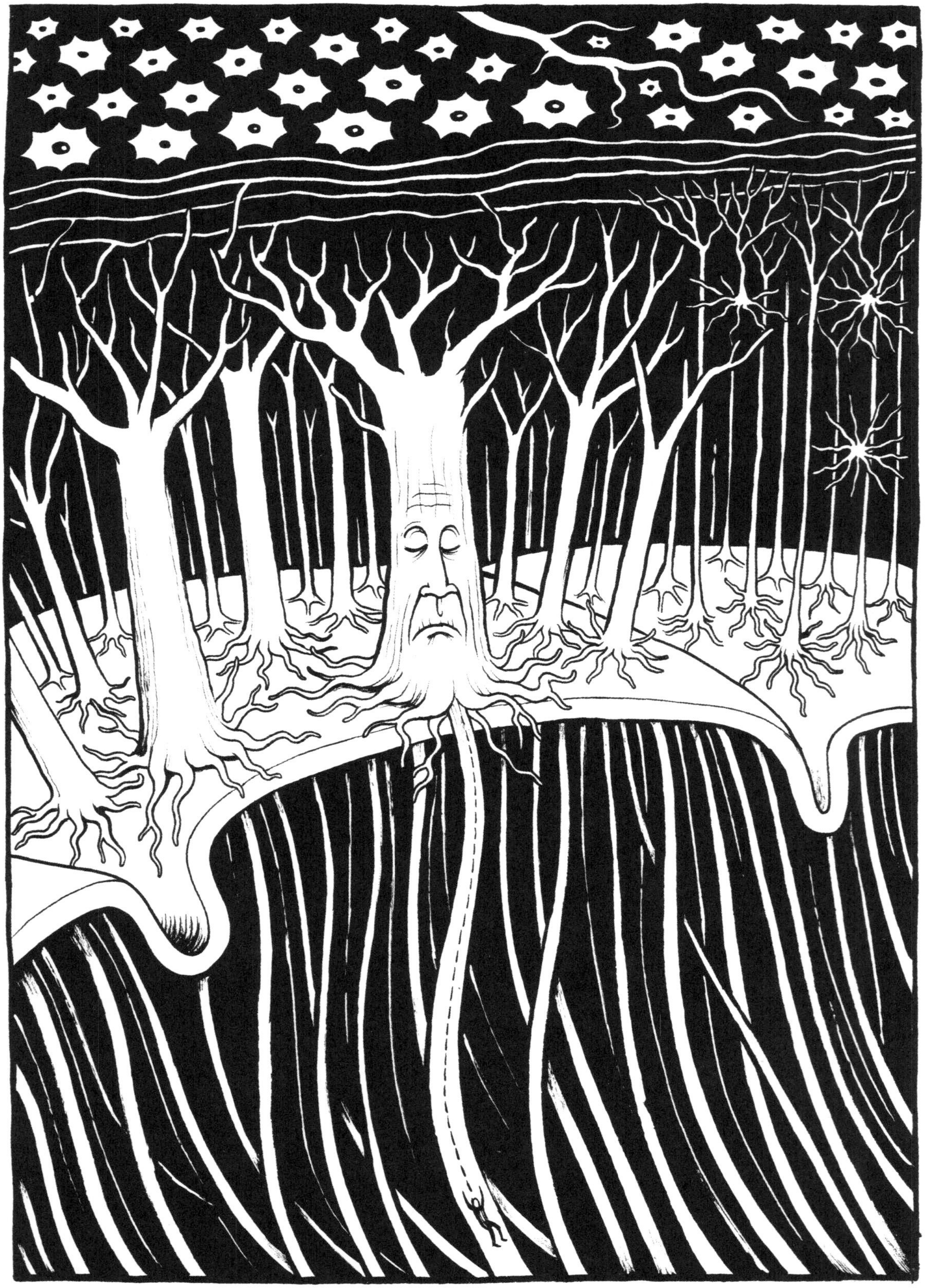

PHARMACOLOGY

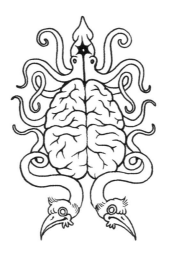

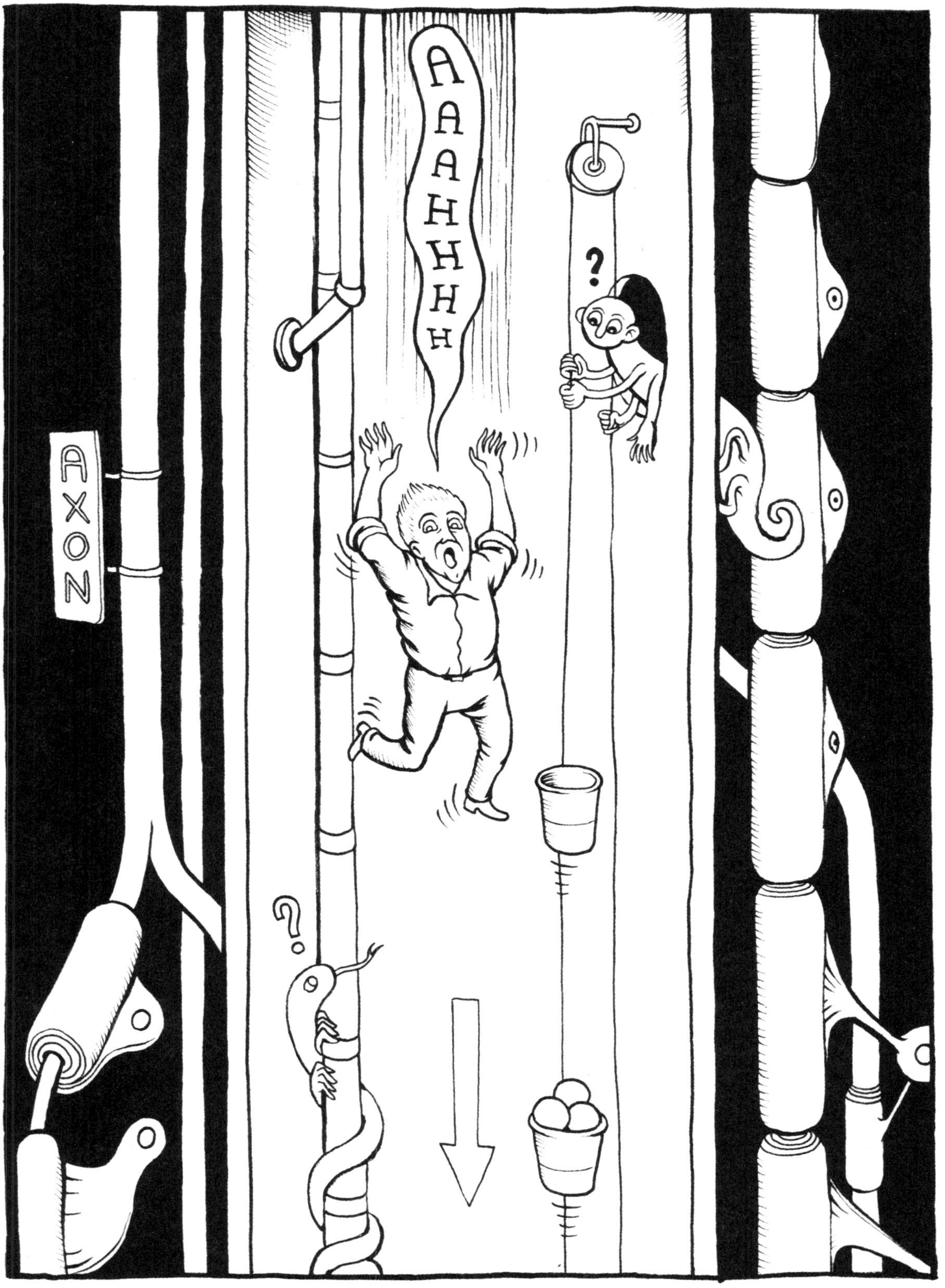
AXON
AAAAHHHH
?
?

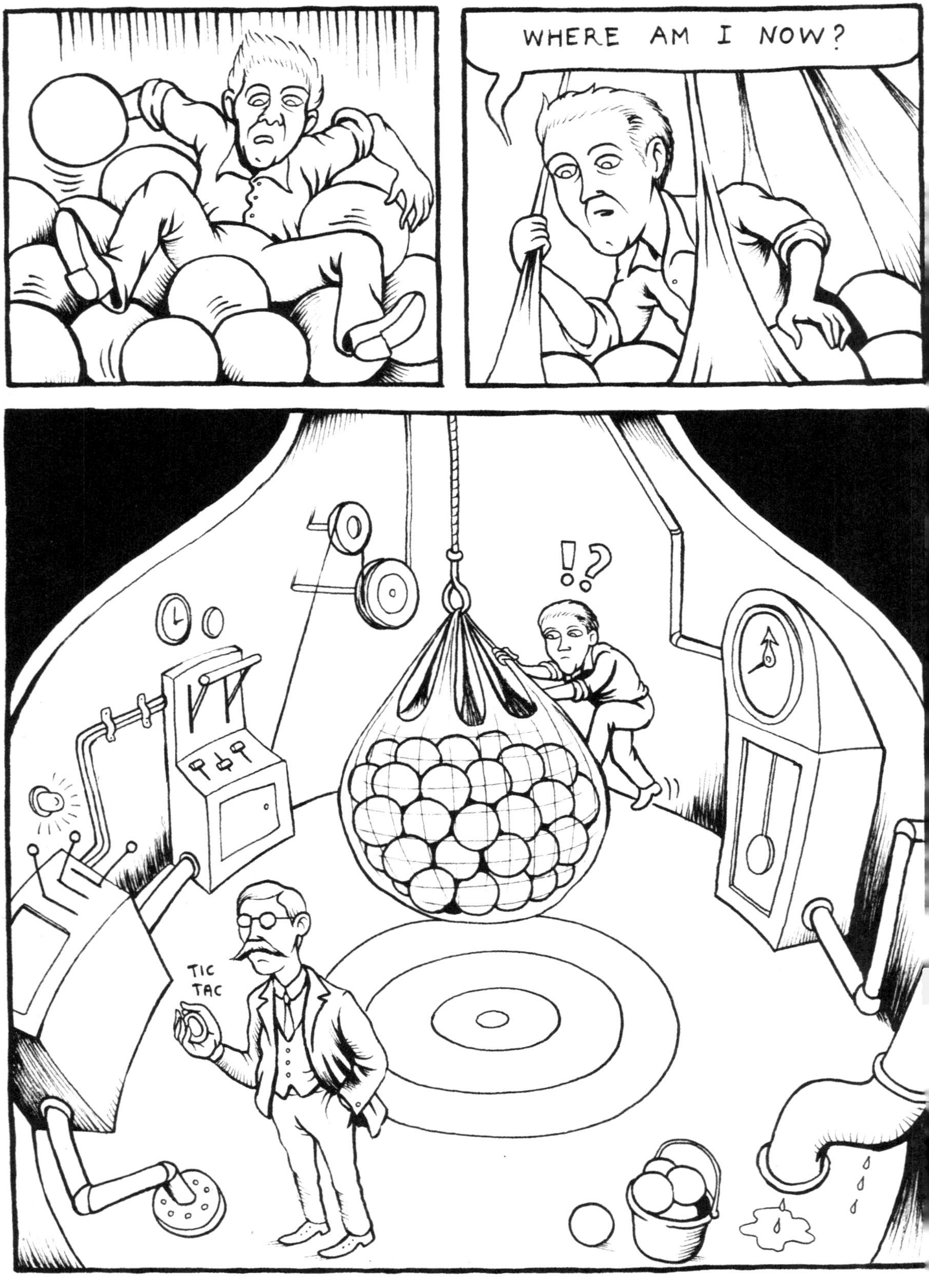
WHERE AM I NOW?
!?
TIC
TAC

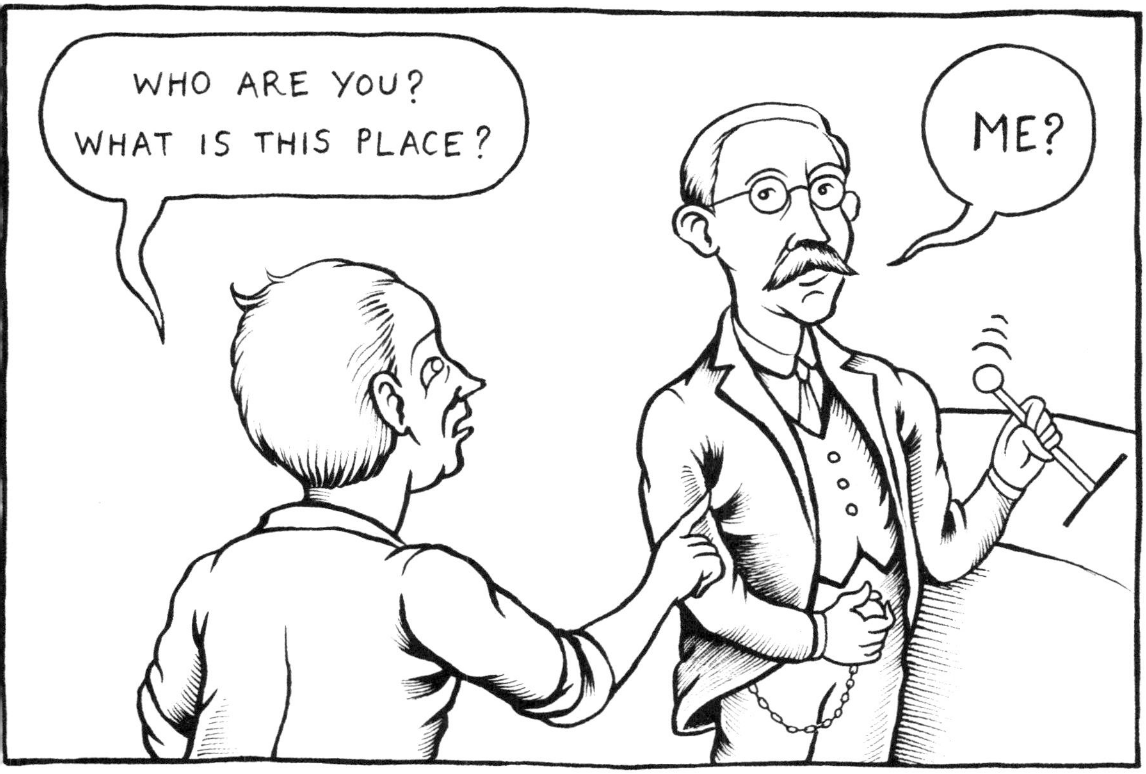
WHO ARE YOU? WHAT IS THIS PLACE?
ME?

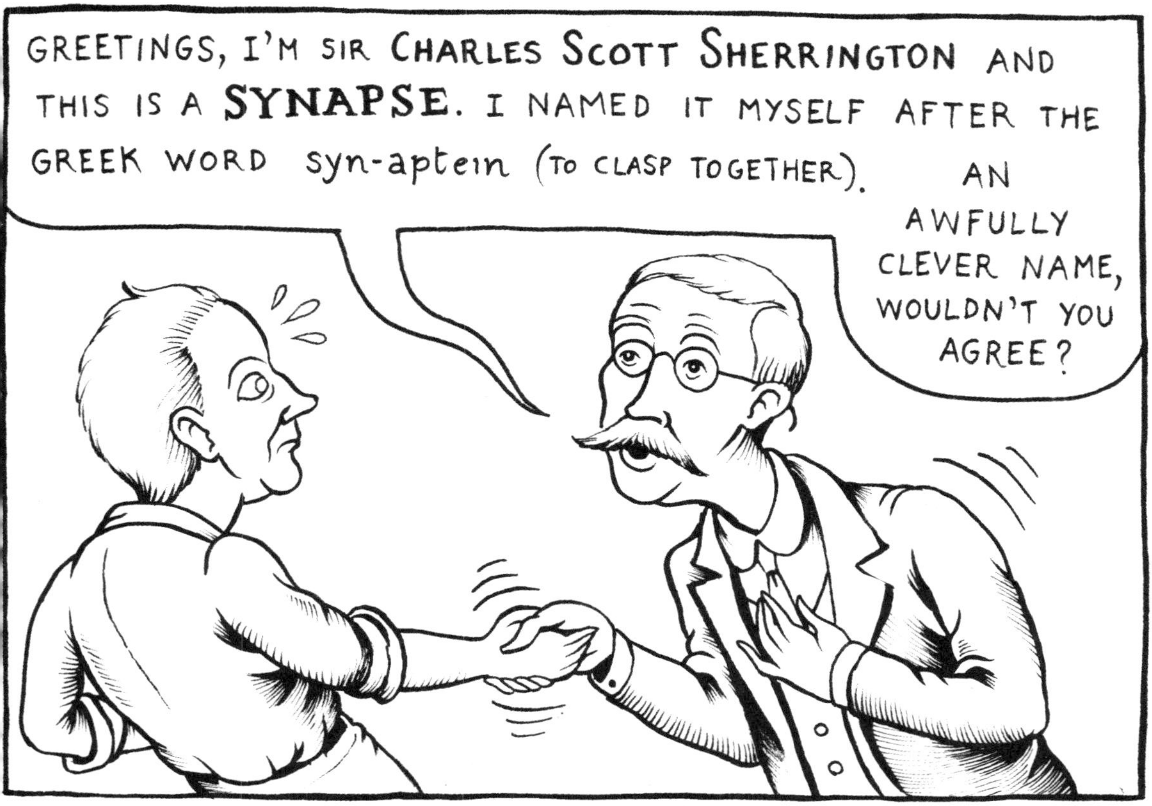
GREETINGS, I'M SIR CHARLES SCOTT SHERRINGTON AND THIS IS A SYNAPSE. I NAMED IT MYSELF AFTER THE GREEK WORD syn-aptein (TO CLASP TOGETHER). AN AWFULLY CLEVER NAME, WOULDN'T YOU AGREE?

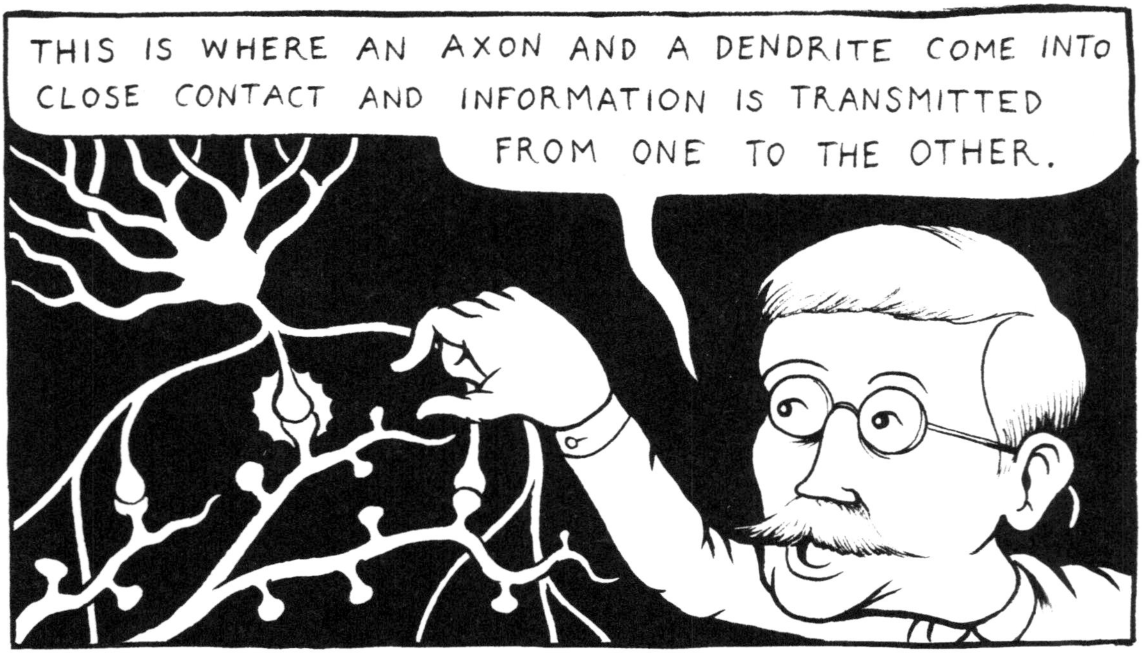
THIS IS WHERE AN AXON AND A DENDRITE COME INTO CLOSE CONTACT AND INFORMATION IS TRANSMITTED FROM ONE TO THE OTHER.

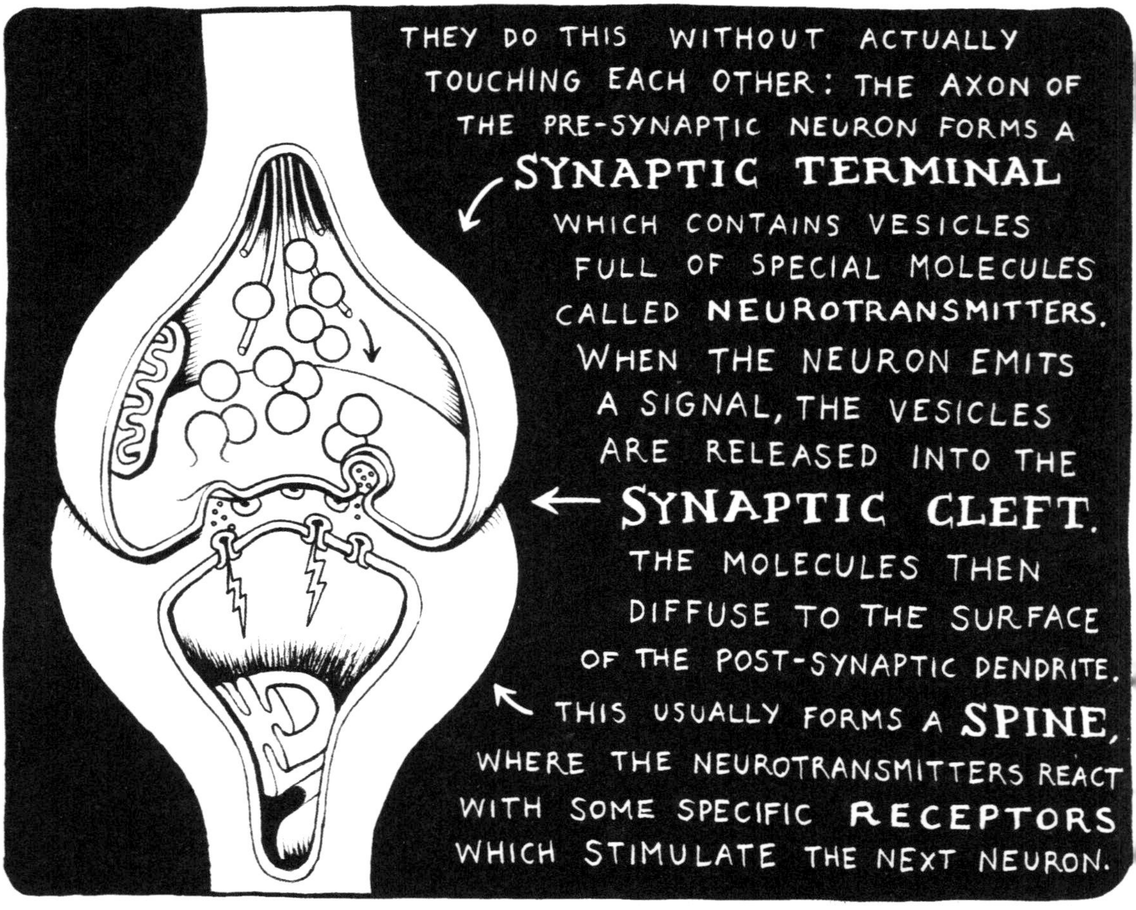
THEY DO THIS WITHOUT ACTUALLY TOUCHING EACH OTHER: THE AXON OF THE PRE-SYNAPTIC NEURON FORMS A
SYNAPTIC TERMINAL
WHICH CONTAINS VESICLES FULL OF SPECIAL MOLECULES CALLED NEUROTRANSMITTERS.
WHEN THE NEURON EMITS A SIGNAL, THE VESICLES ARE RELEASED INTO THE
SYNAPTIC CLEFT.
THE MOLECULES THEN DIFFUSE TO THE SURFACE OF THE POST-SYNAPTIC DENDRITE.
THIS USUALLY FORMS A SPINE, WHERE THE NEUROTRANSMITTERS REACT WITH SOME SPECIFIC RECEPTORS WHICH STIMULATE THE NEXT NEURON.

SYNAPTIC TRANSMISSION
HAS TWO GREAT
ADVANTAGES:

THE FIRST IS THAT THE
SAME SIGNAL CAN HAVE
DIFFERENT MEANINGS
DEPENDING ON THE
COMBINATION OF
MOLECULES AND RECEPTORS
PRESENT IN THE SYNAPSE.

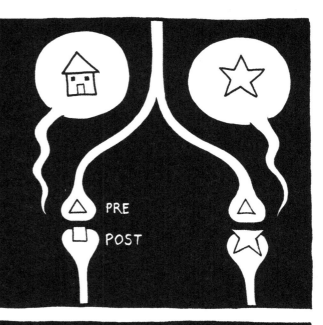

SECONDLY, WHEN A NEURON
EMITS A SIGNAL THIS IS
TRANSMITTED TO ALL ITS
SYNAPTIC TERMINALS, BUT
EACH NEURON REQUIRES THE
ACTIVATION OF MANY
SYNAPSES IN ORDER TO
GENERATE A NEW SIGNAL.

NOT ALL OF THEM WILL
BE ACTIVE AT THE SAME
TIME AND THIS IS WHAT
UNDERLIES
COMPUTATION
IN THE BRAIN.

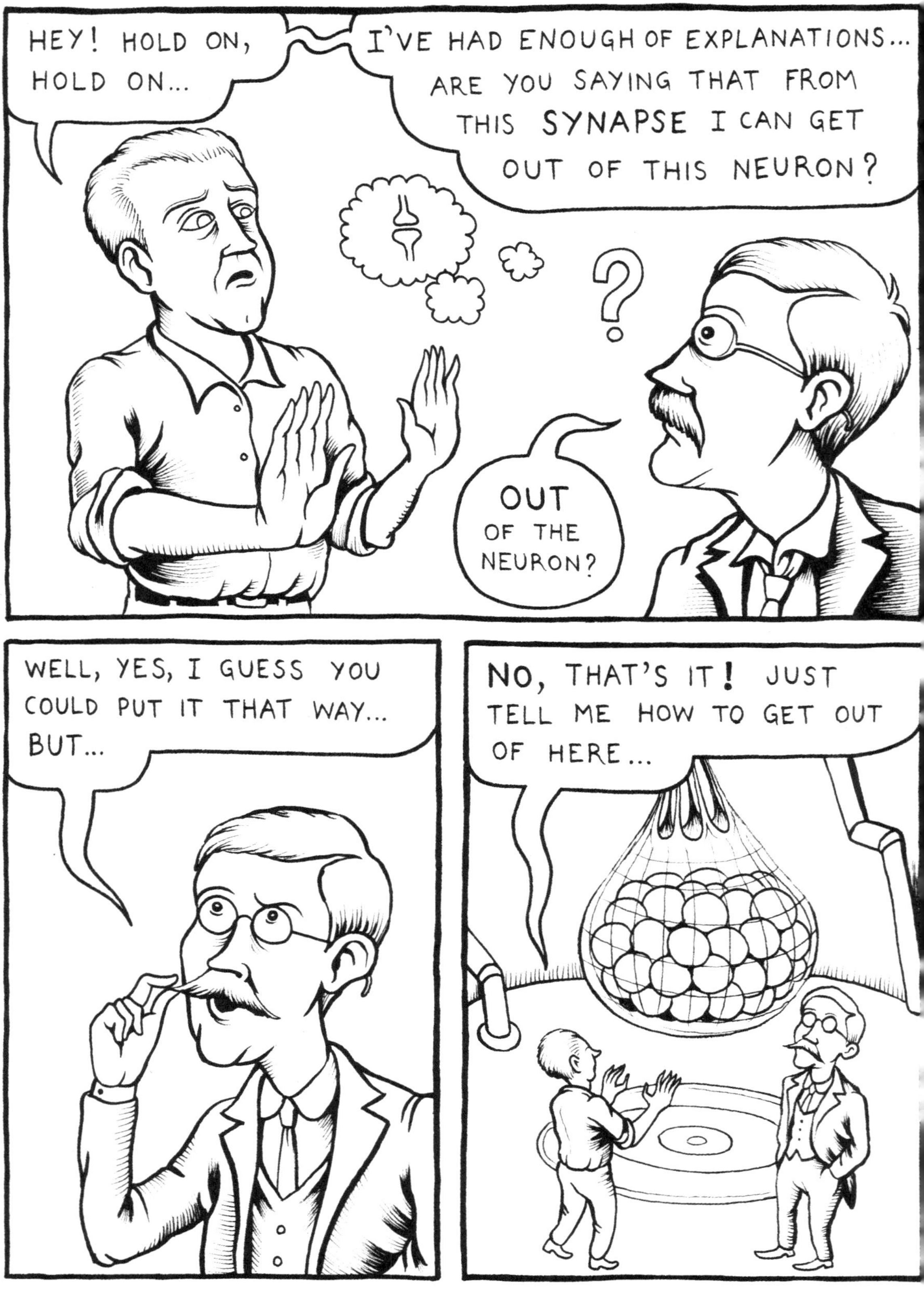
HEY! HOLD ON, HOLD ON...
I'VE HAD ENOUGH OF EXPLANATIONS... ARE YOU SAYING THAT FROM THIS SYNAPSE I CAN GET OUT OF THIS NEURON?
OUT OF THE NEURON?
WELL, YES, I GUESS YOU COULD PUT IT THAT WAY... BUT...
NO, THAT'S IT! JUST TELL ME HOW TO GET OUT OF HERE...

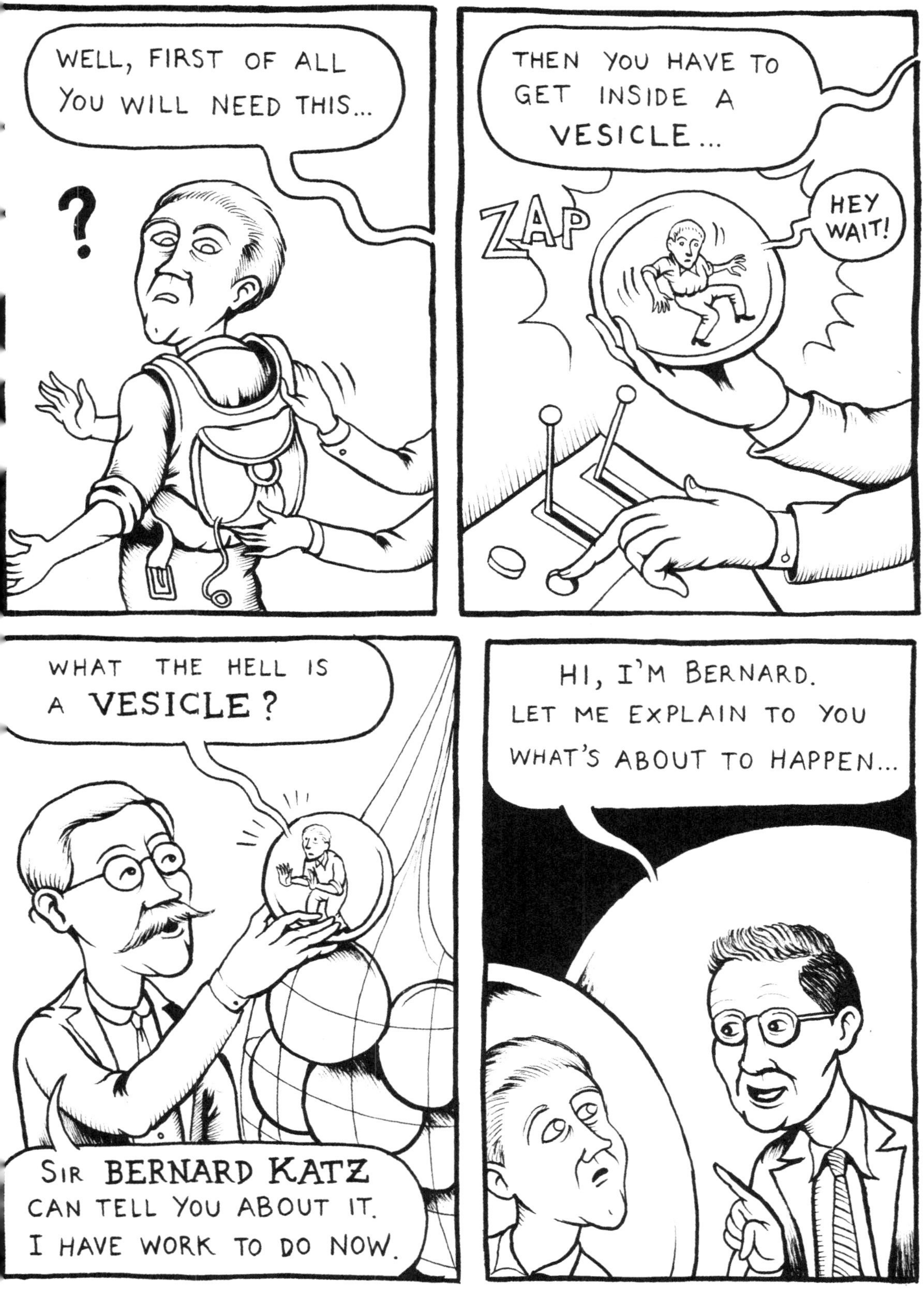
WELL, FIRST OF ALL YOU WILL NEED THIS...
?
THEN YOU HAVE TO GET INSIDE A VESICLE...
ZAP
HEY WAIT!
WHAT THE HELL IS A VESICLE?
SIR BERNARD KATZ CAN TELL YOU ABOUT IT. I HAVE WORK TO DO NOW.
HI, I'M BERNARD. LET ME EXPLAIN TO YOU WHAT'S ABOUT TO HAPPEN...

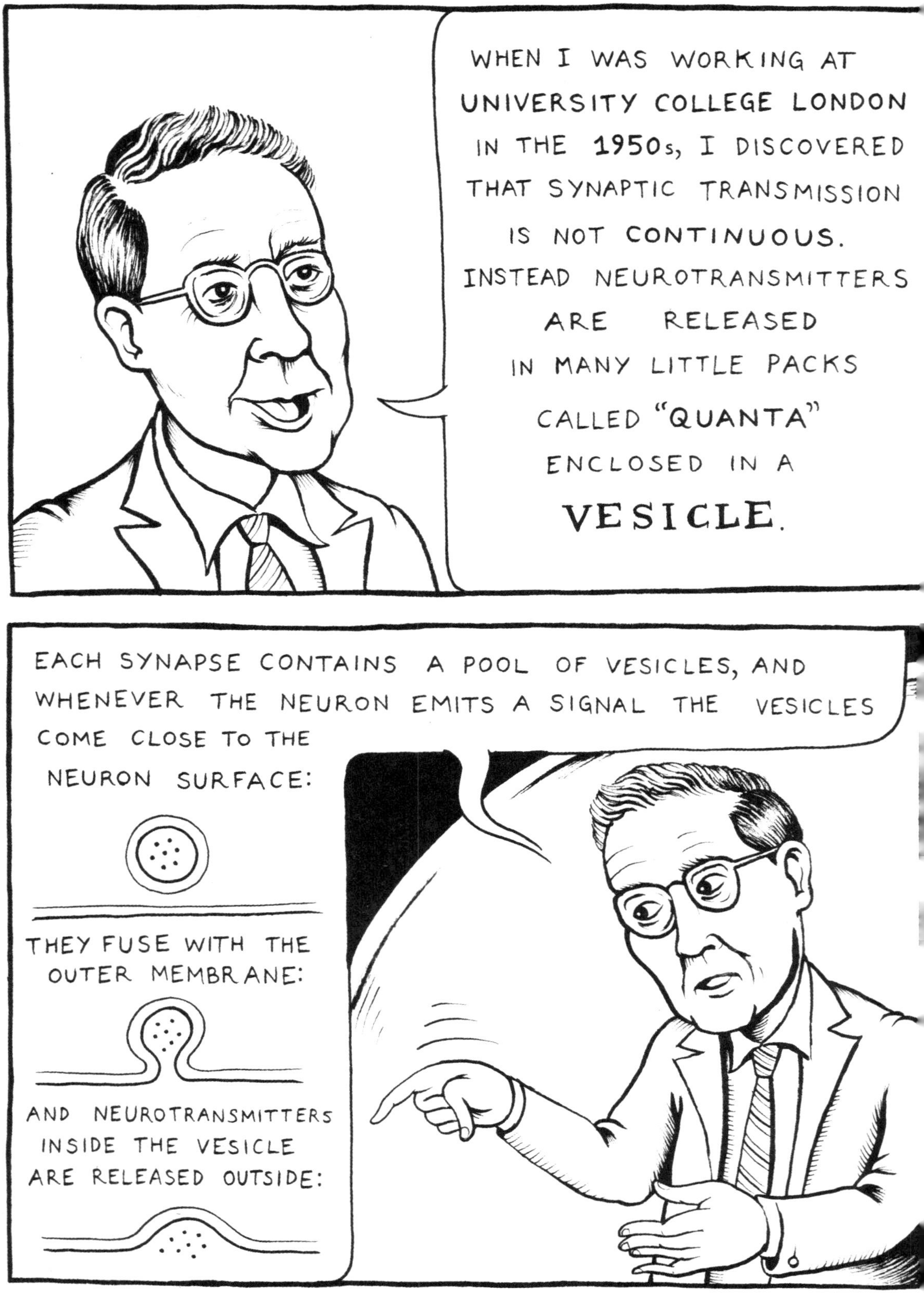
WHEN I WAS WORKING AT UNIVERSITY COLLEGE LONDON IN THE 1950s, I DISCOVERED THAT SYNAPTIC TRANSMISSION IS NOT CONTINUOUS. INSTEAD NEUROTRANSMITTERS ARE RELEASED IN MANY LITTLE PACKS CALLED "QUANTA" ENCLOSED IN A VESICLE.
EACH SYNAPSE CONTAINS A POOL OF VESICLES, AND WHENEVER THE NEURON EMITS A SIGNAL THE VESICLES COME CLOSE TO THE NEURON SURFACE:
THEY FUSE WITH THE OUTER MEMBRANE:
AND NEUROTRANSMITTERS INSIDE THE VESICLE ARE RELEASED OUTSIDE:

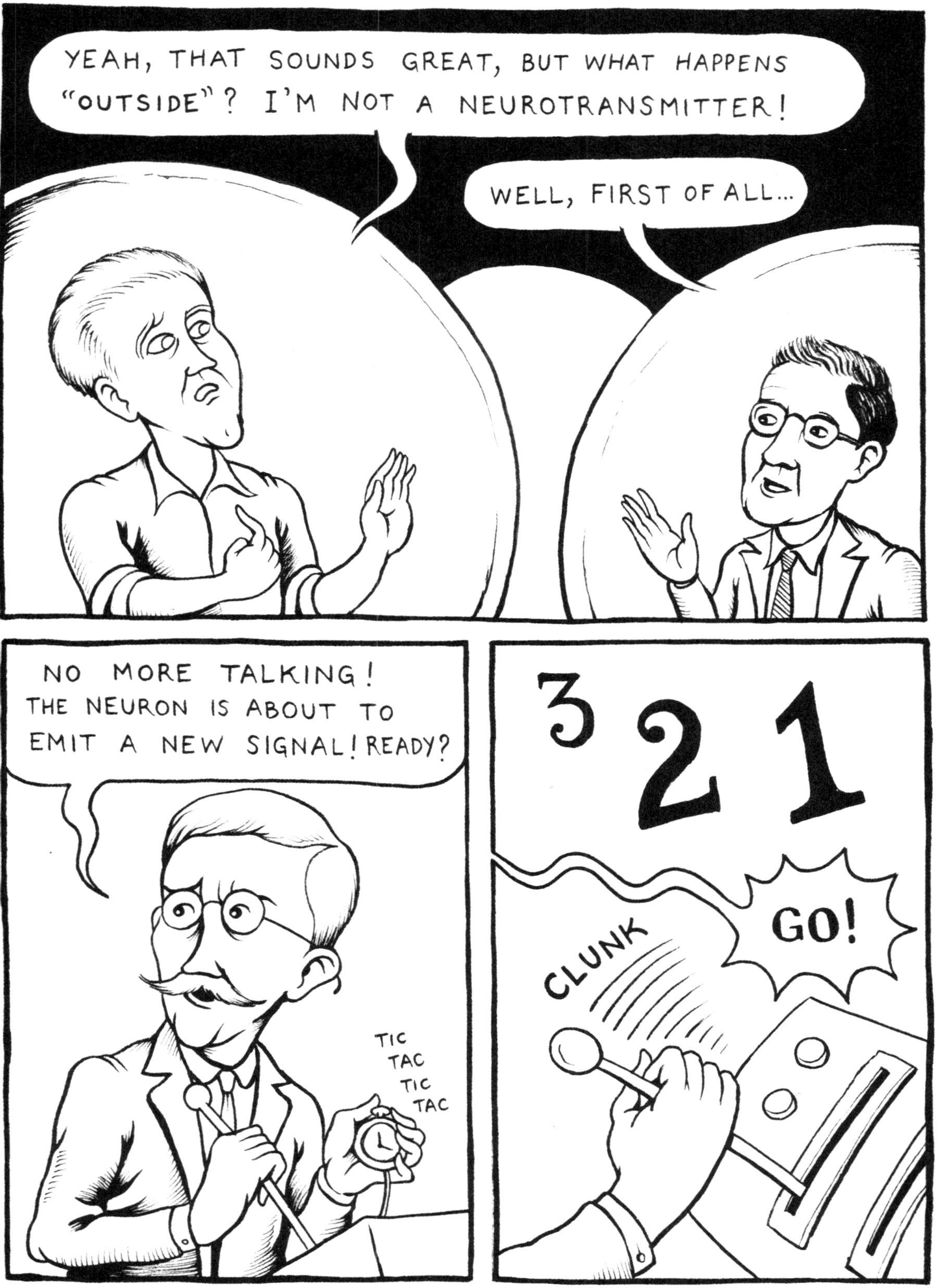
YEAH, THAT SOUNDS GREAT, BUT WHAT HAPPENS "OUTSIDE"? I'M NOT A NEUROTRANSMITTER!
WELL, FIRST OF ALL...
NO MORE TALKING! THE NEURON IS ABOUT TO EMIT A NEW SIGNAL! READY?
TIC TAC TIC TAC
3 2 1
GO!
CLUNK

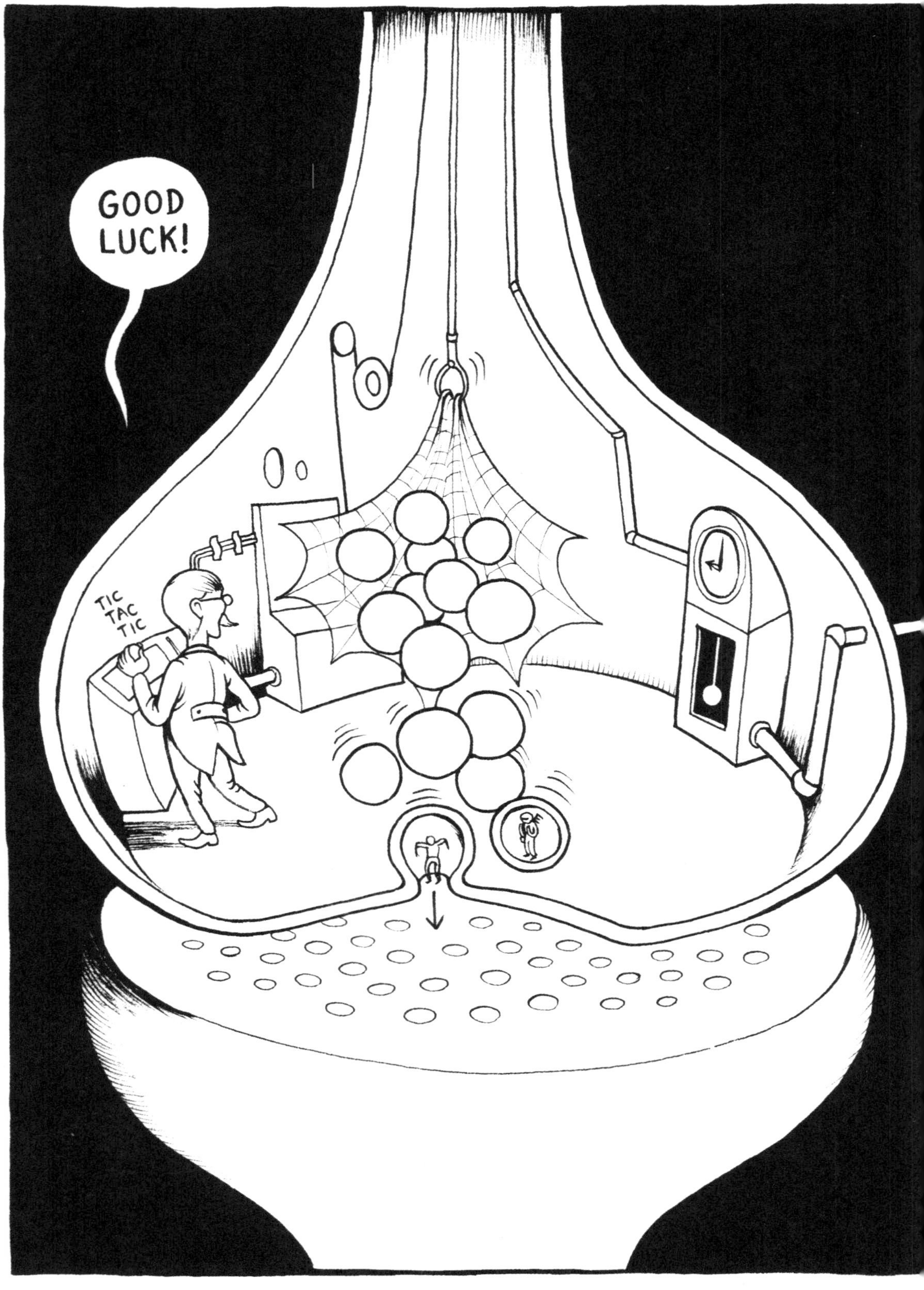
GOOD LUCK!
TIC TAC TIC

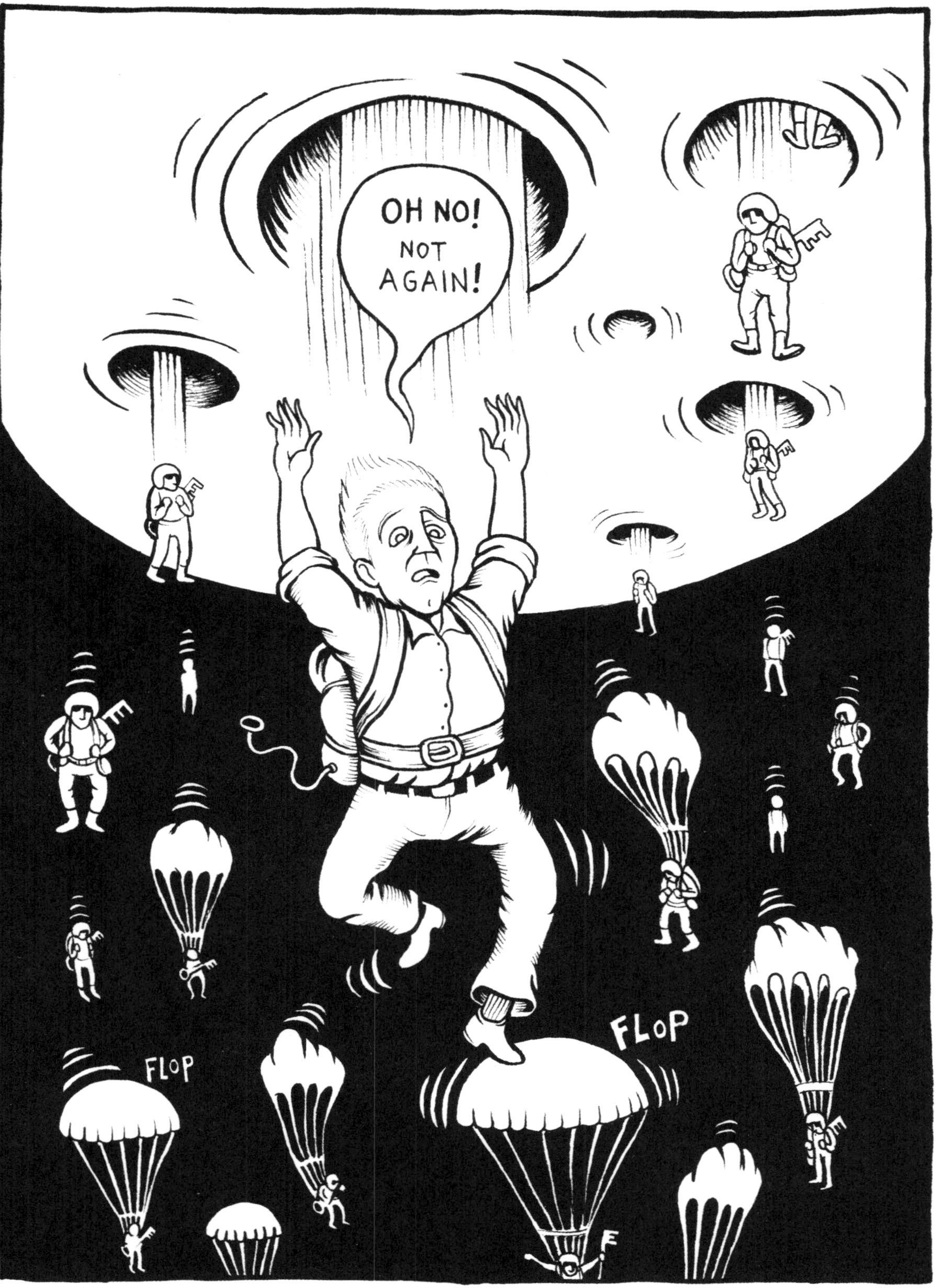
OH NO! NOT AGAIN!
FLOP
FLOP

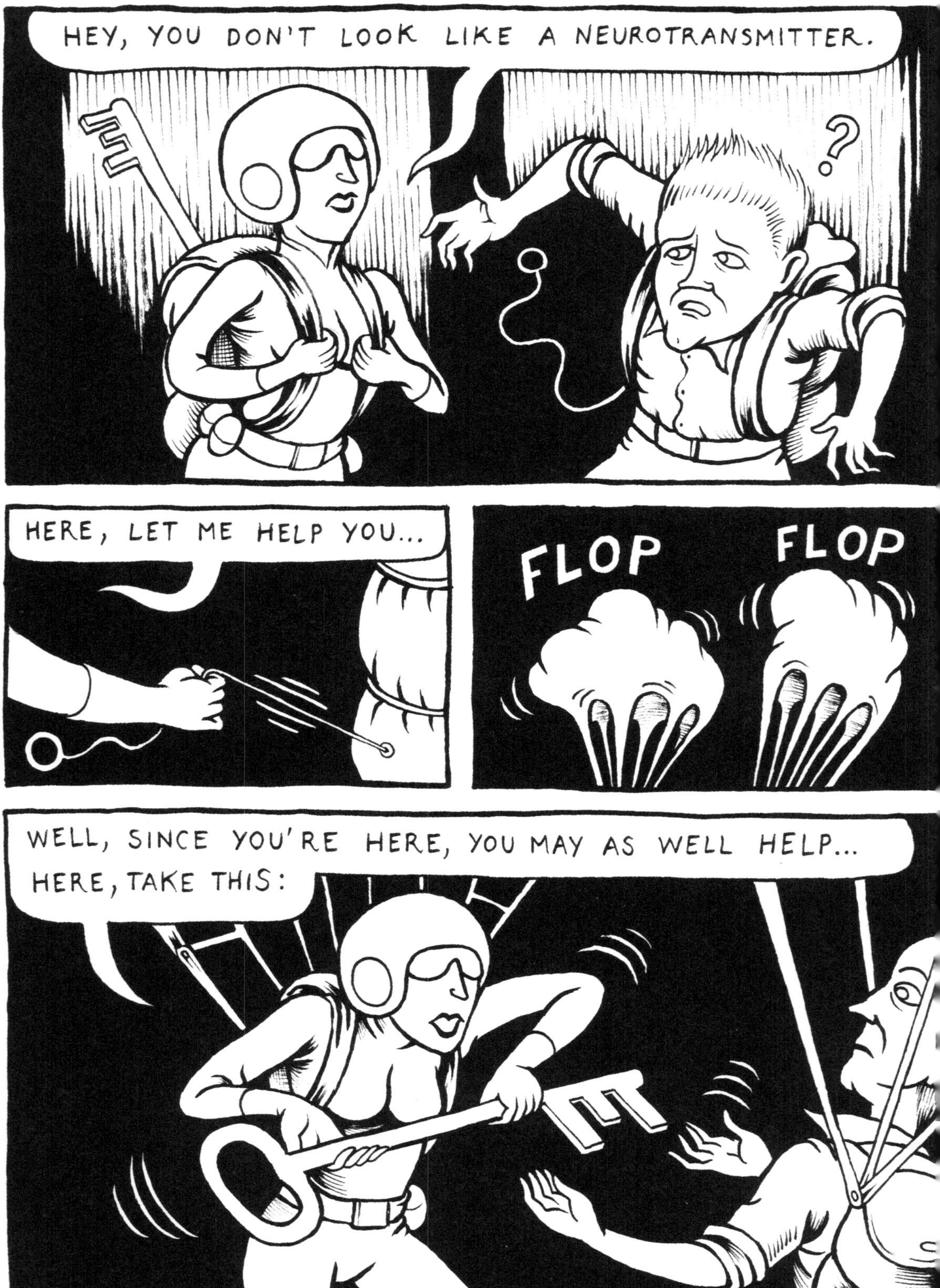
HEY, YOU DON'T LOOK LIKE A NEUROTRANSMITTER.
?
HERE, LET ME HELP YOU...
FLOP
FLOP
WELL, SINCE YOU'RE HERE, YOU MAY AS WELL HELP... HERE, TAKE THIS:

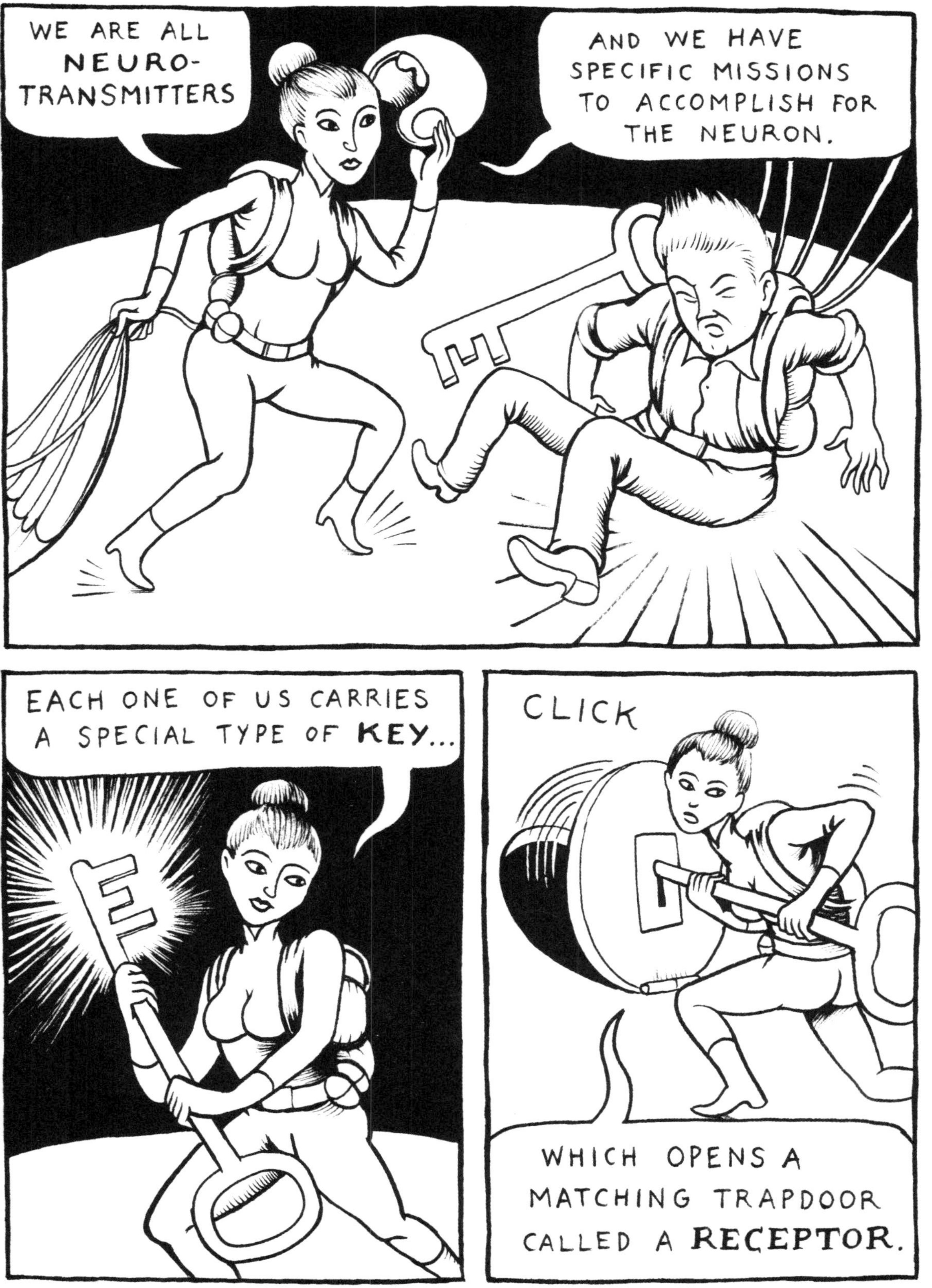
WE ARE ALL NEURO-TRANSMITTERS
AND WE HAVE SPECIFIC MISSIONS TO ACCOMPLISH FOR THE NEURON.
EACH ONE OF US CARRIES A SPECIAL TYPE OF KEY...
CLICK
WHICH OPENS A MATCHING TRAPDOOR CALLED A RECEPTOR.

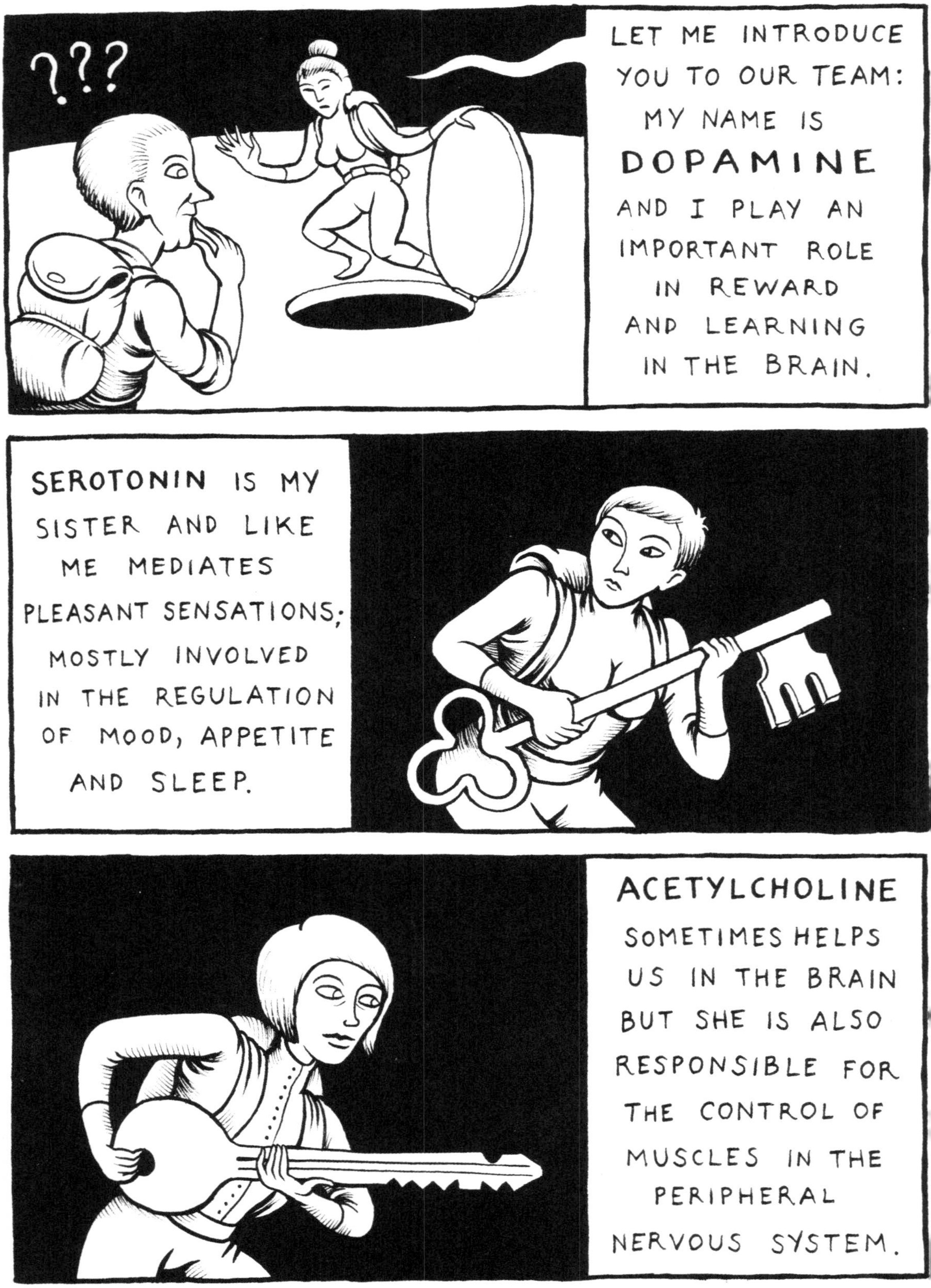
???
LET ME INTRODUCE YOU TO OUR TEAM: MY NAME IS DOPAMINE AND I PLAY AN IMPORTANT ROLE IN REWARD AND LEARNING IN THE BRAIN.
SEROTONIN IS MY SISTER AND LIKE ME MEDIATES PLEASANT SENSATIONS; MOSTLY INVOLVED IN THE REGULATION OF MOOD, APPETITE AND SLEEP.
ACETYLCHOLINE SOMETIMES HELPS US IN THE BRAIN BUT SHE IS ALSO RESPONSIBLE FOR THE CONTROL OF MUSCLES IN THE PERIPHERAL NERVOUS SYSTEM.

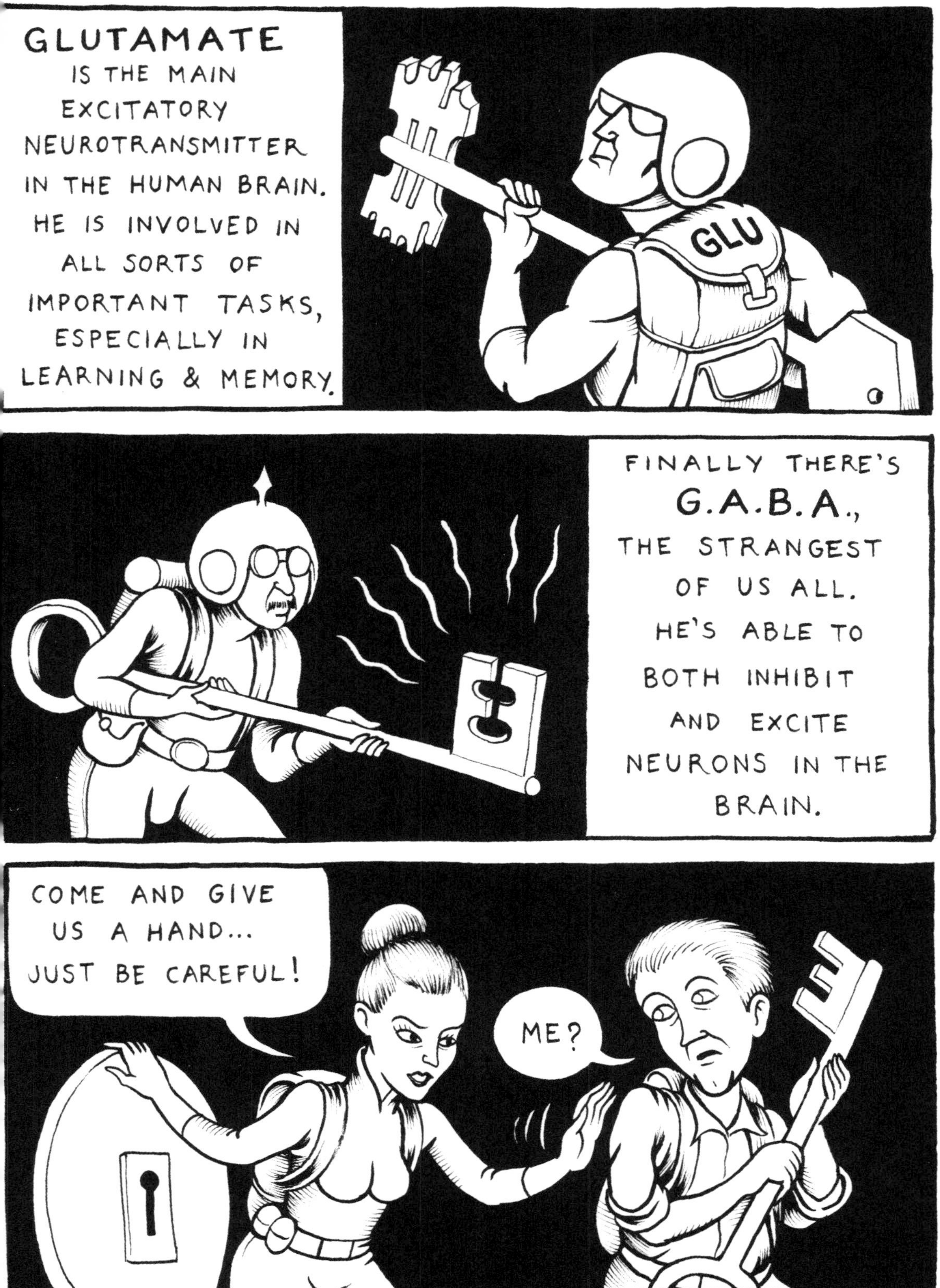
GLUTAMATE IS THE MAIN EXCITATORY NEUROTRANSMITTER IN THE HUMAN BRAIN. HE IS INVOLVED IN ALL SORTS OF IMPORTANT TASKS, ESPECIALLY IN LEARNING & MEMORY.
GLU
FINALLY THERE'S G.A.B.A., THE STRANGEST OF US ALL. HE'S ABLE TO BOTH INHIBIT AND EXCITE NEURONS IN THE BRAIN.
COME AND GIVE US A HAND... JUST BE CAREFUL!
ME?

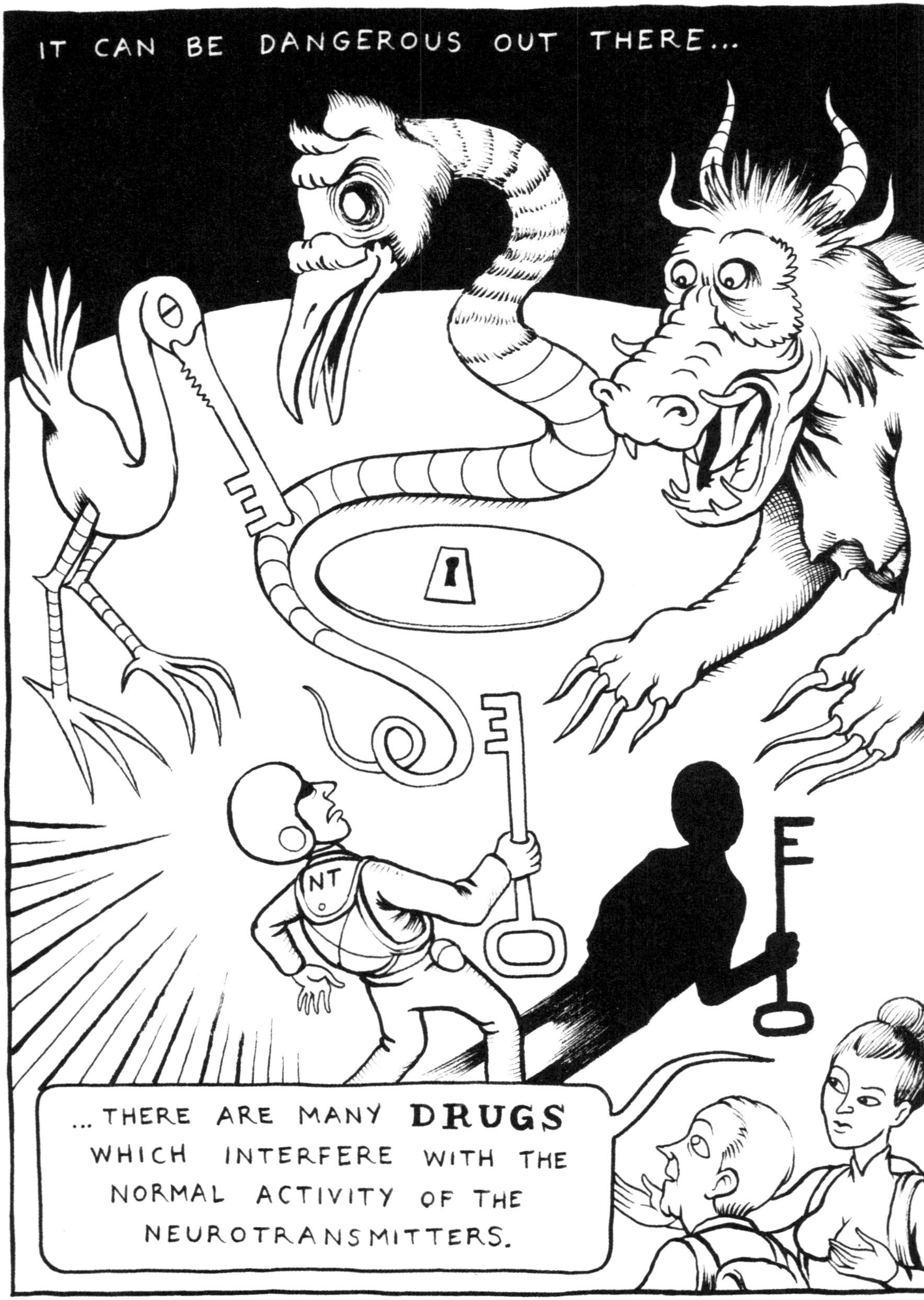
IT CAN BE DANGEROUS OUT THERE...
NT
...THERE ARE MANY DRUGS WHICH INTERFERE WITH THE NORMAL ACTIVITY OF THE NEUROTRANSMITTERS.

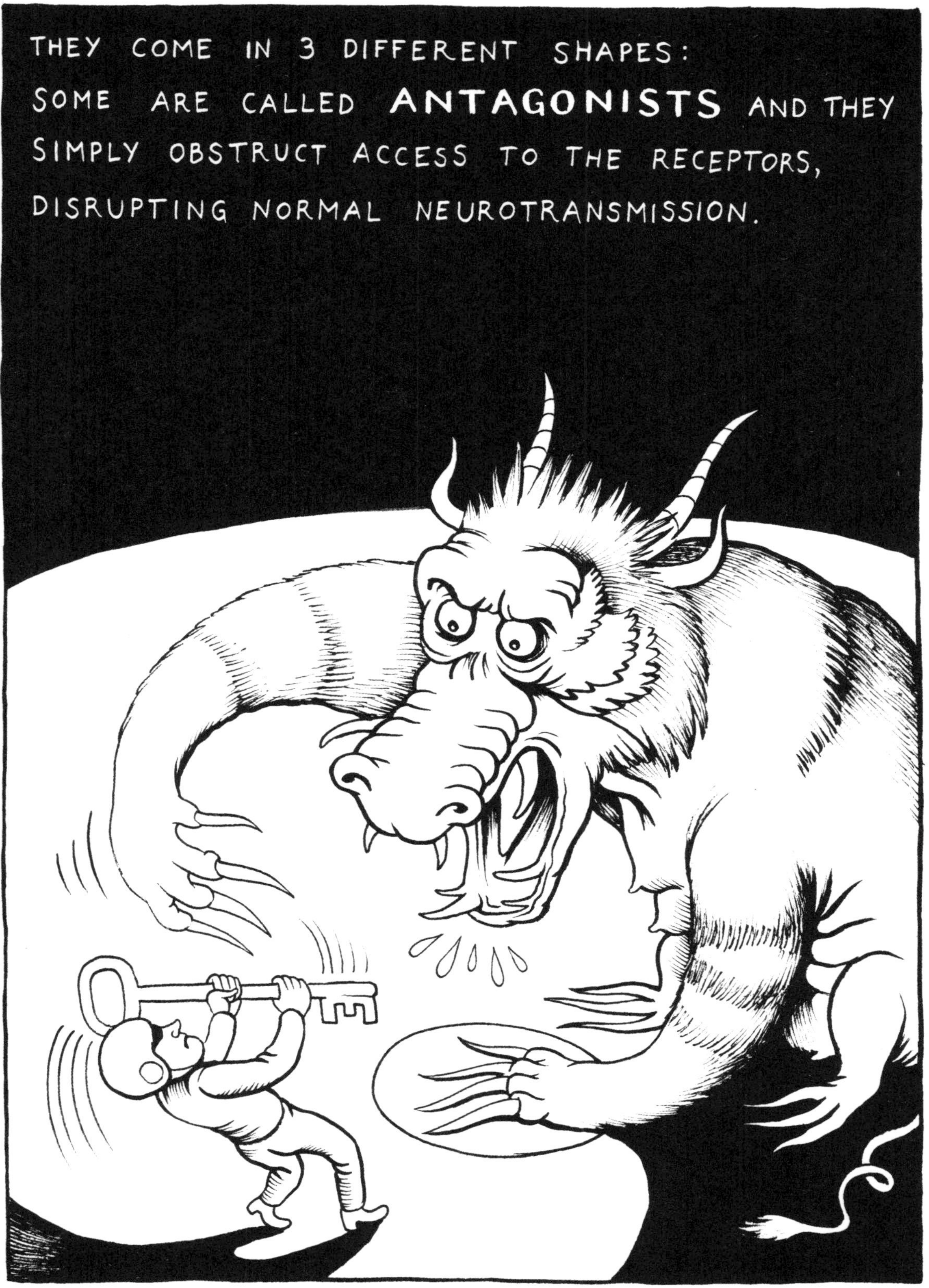
THEY COME IN 3 DIFFERENT SHAPES:
SOME ARE CALLED **ANTAGONISTS** AND THEY SIMPLY OBSTRUCT ACCESS TO THE RECEPTORS, DISRUPTING NORMAL NEUROTRANSMISSION.

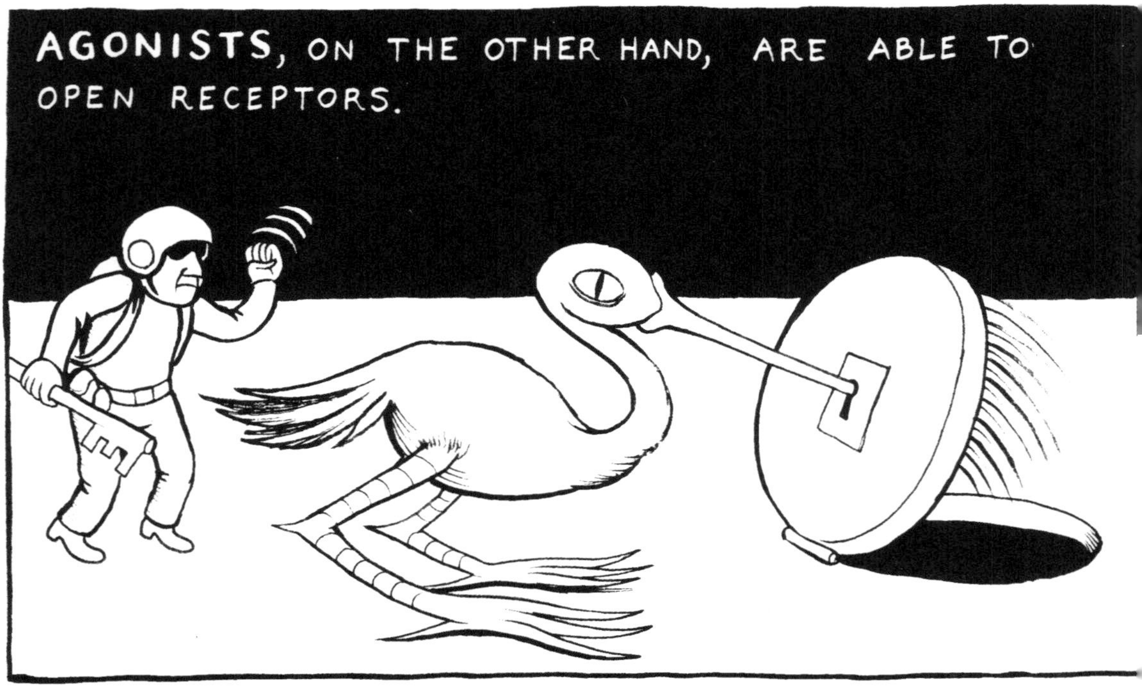
AGONISTS, ON THE OTHER HAND, ARE ABLE TO OPEN RECEPTORS.

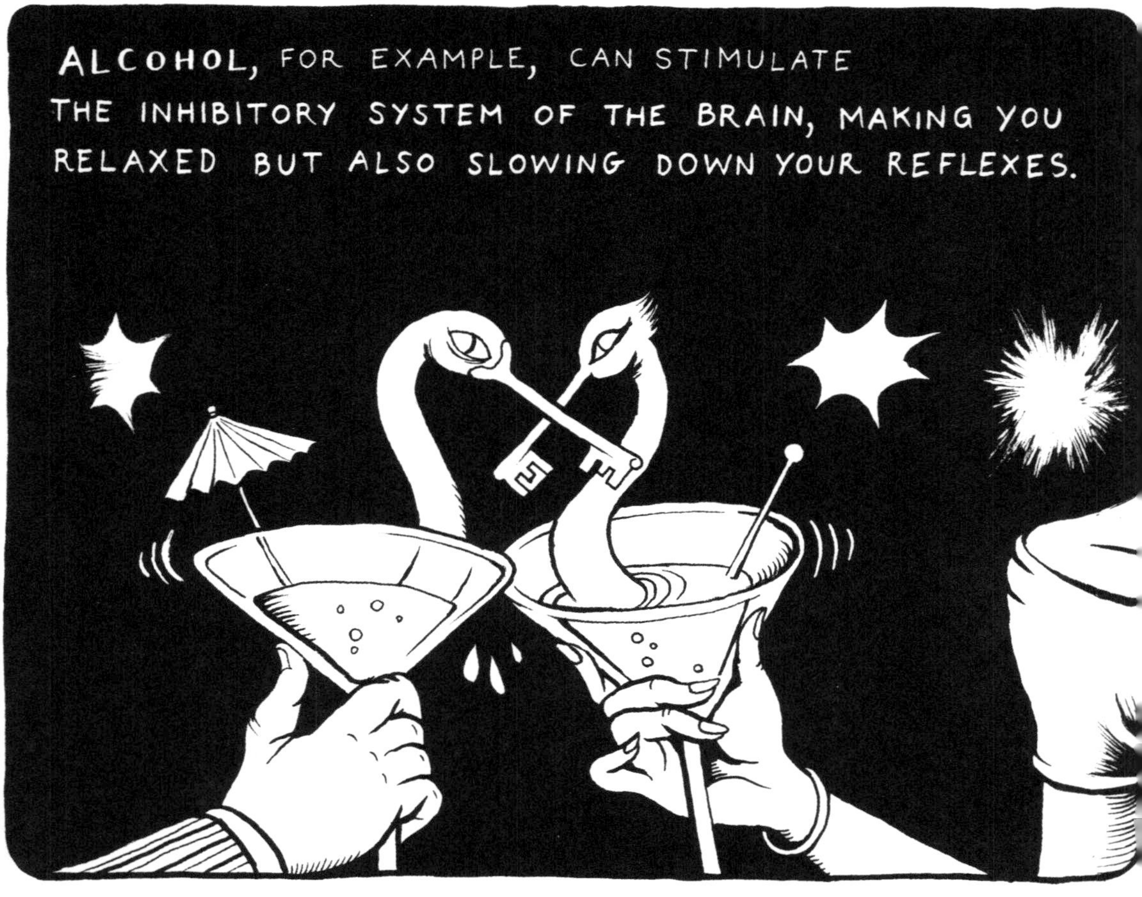
ALCOHOL, FOR EXAMPLE, CAN STIMULATE THE INHIBITORY SYSTEM OF THE BRAIN, MAKING YOU RELAXED BUT ALSO SLOWING DOWN YOUR REFLEXES.

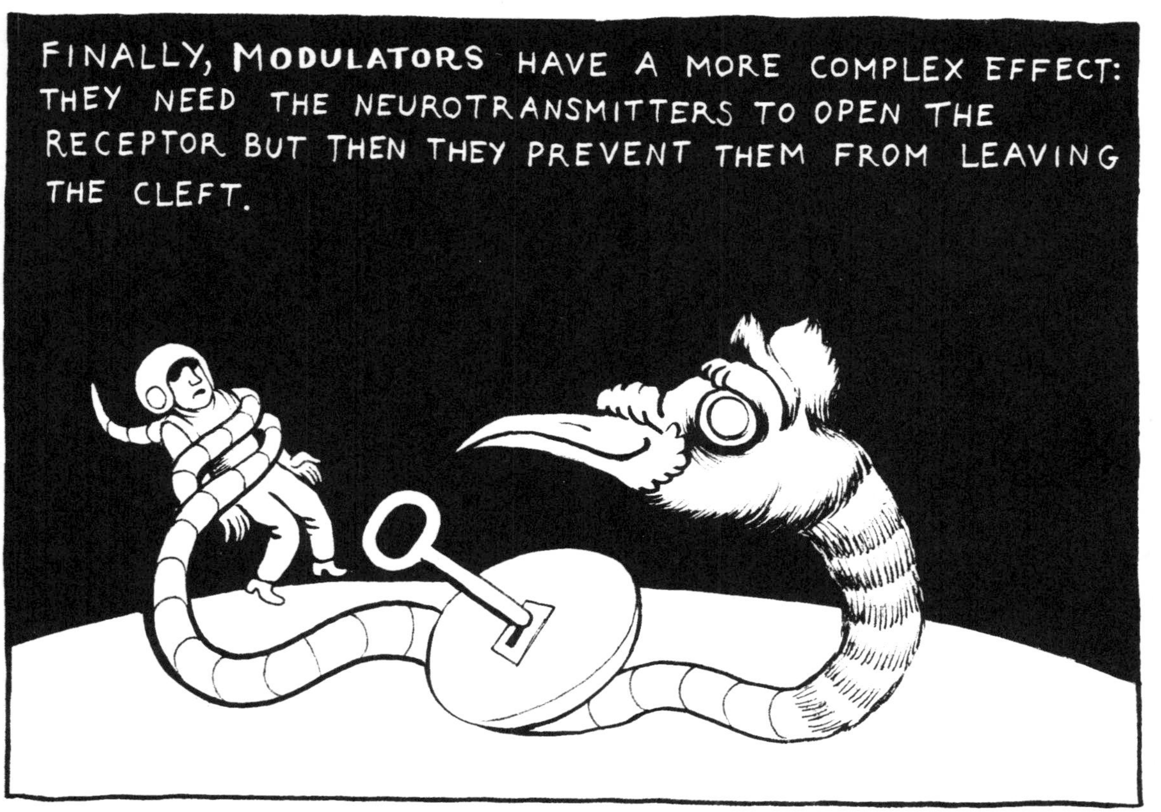
FINALLY, MODULATORS HAVE A MORE COMPLEX EFFECT: THEY NEED THE NEUROTRANSMITTERS TO OPEN THE RECEPTOR BUT THEN THEY PREVENT THEM FROM LEAVING THE CLEFT.

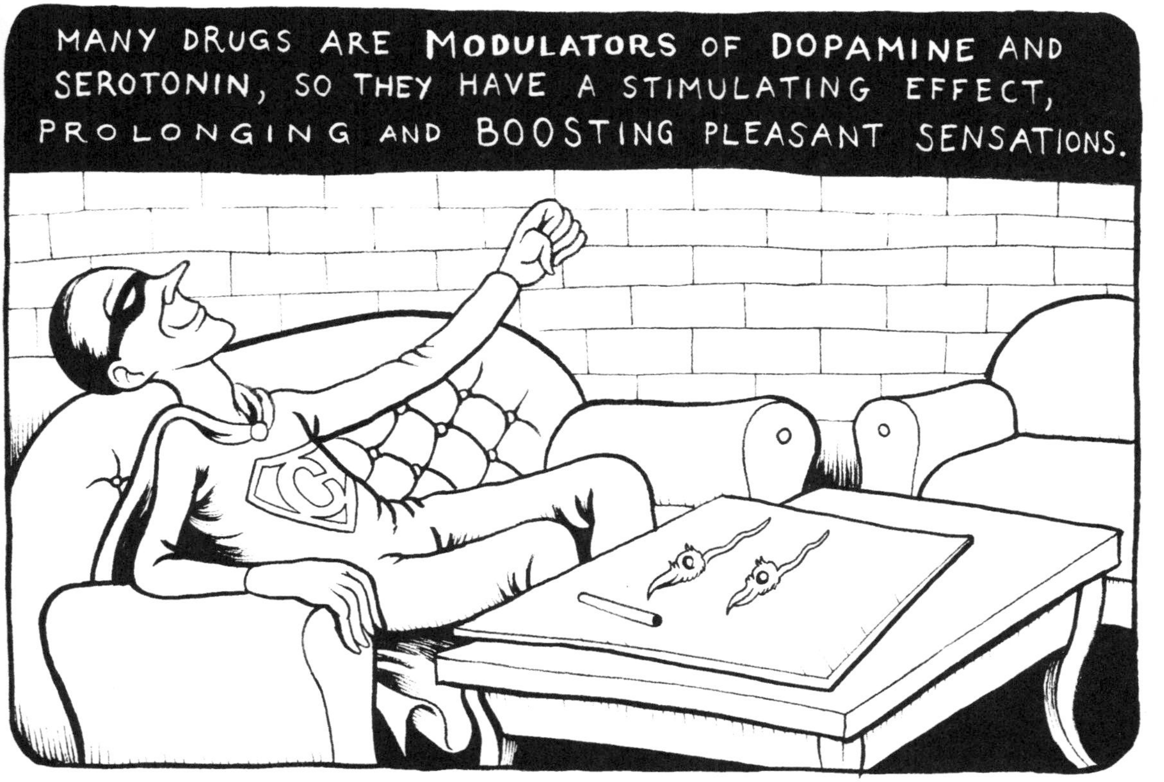
MANY DRUGS ARE MODULATORS OF DOPAMINE AND SEROTONIN, SO THEY HAVE A STIMULATING EFFECT, PROLONGING AND BOOSTING PLEASANT SENSATIONS.

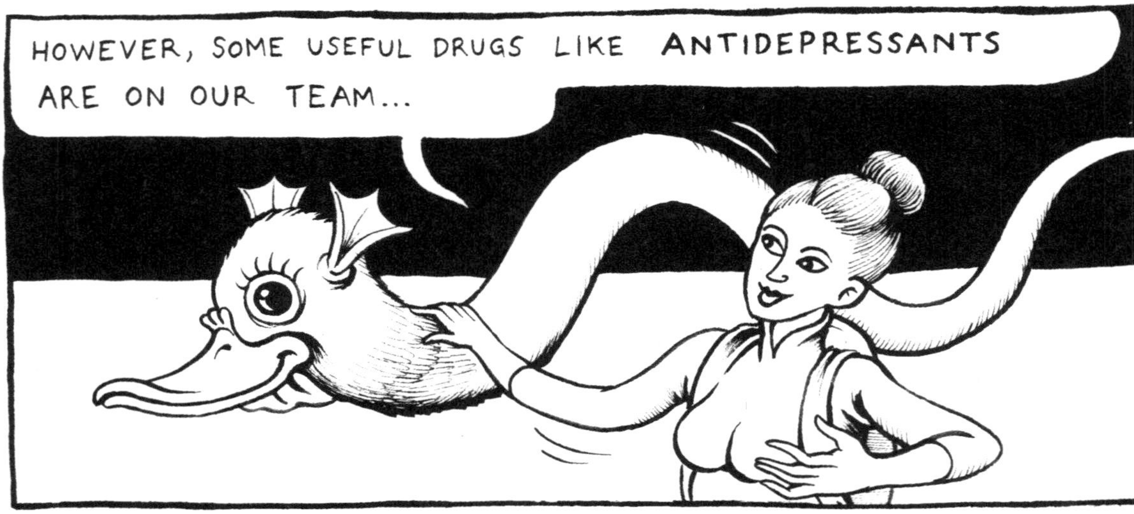
HOWEVER, SOME USEFUL DRUGS LIKE ANTIDEPRESSANTS ARE ON OUR TEAM...

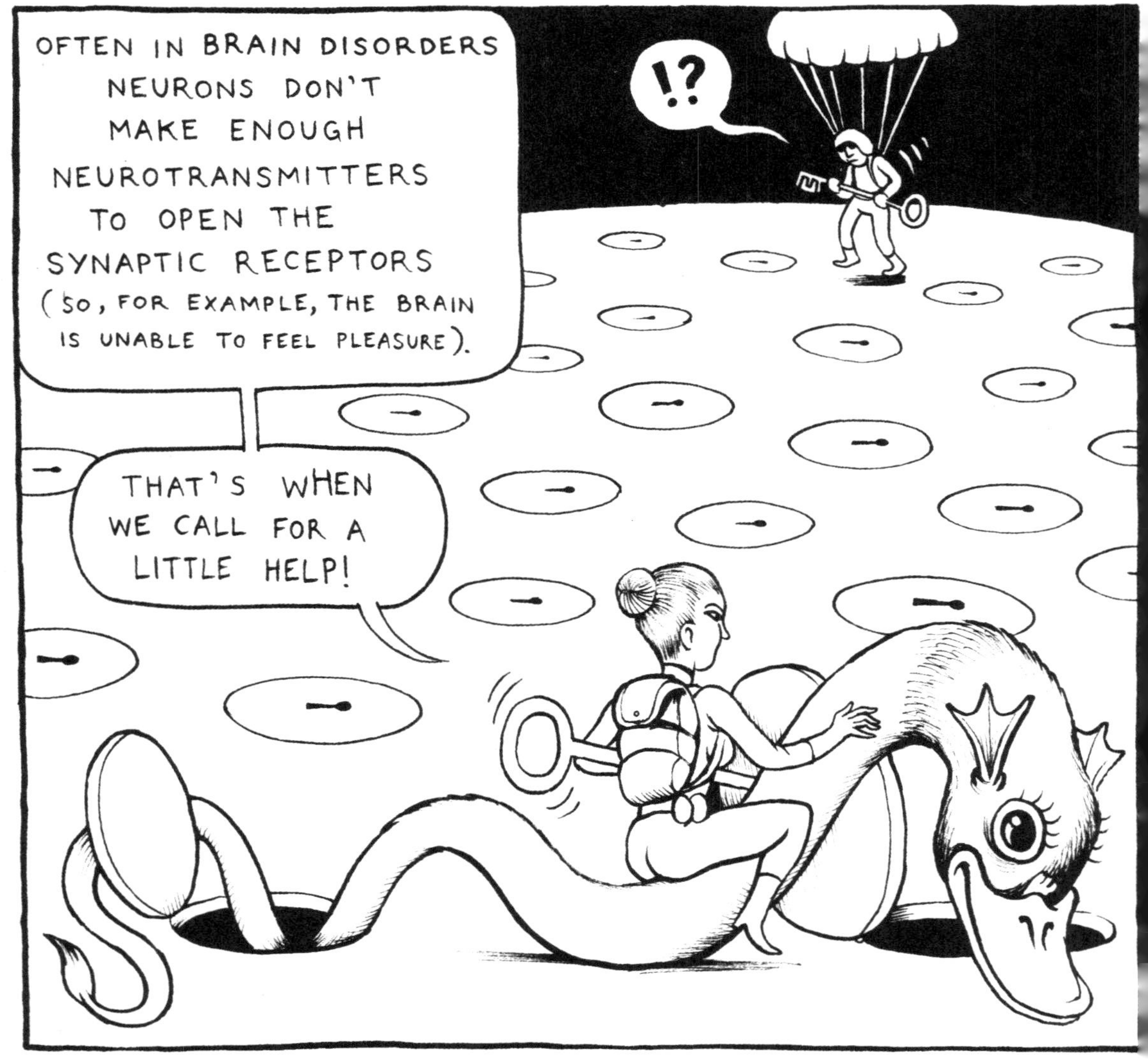
OFTEN IN BRAIN DISORDERS NEURONS DON'T MAKE ENOUGH NEUROTRANSMITTERS TO OPEN THE SYNAPTIC RECEPTORS (SO, FOR EXAMPLE, THE BRAIN IS UNABLE TO FEEL PLEASURE).
THAT'S WHEN WE CALL FOR A LITTLE HELP!
!?

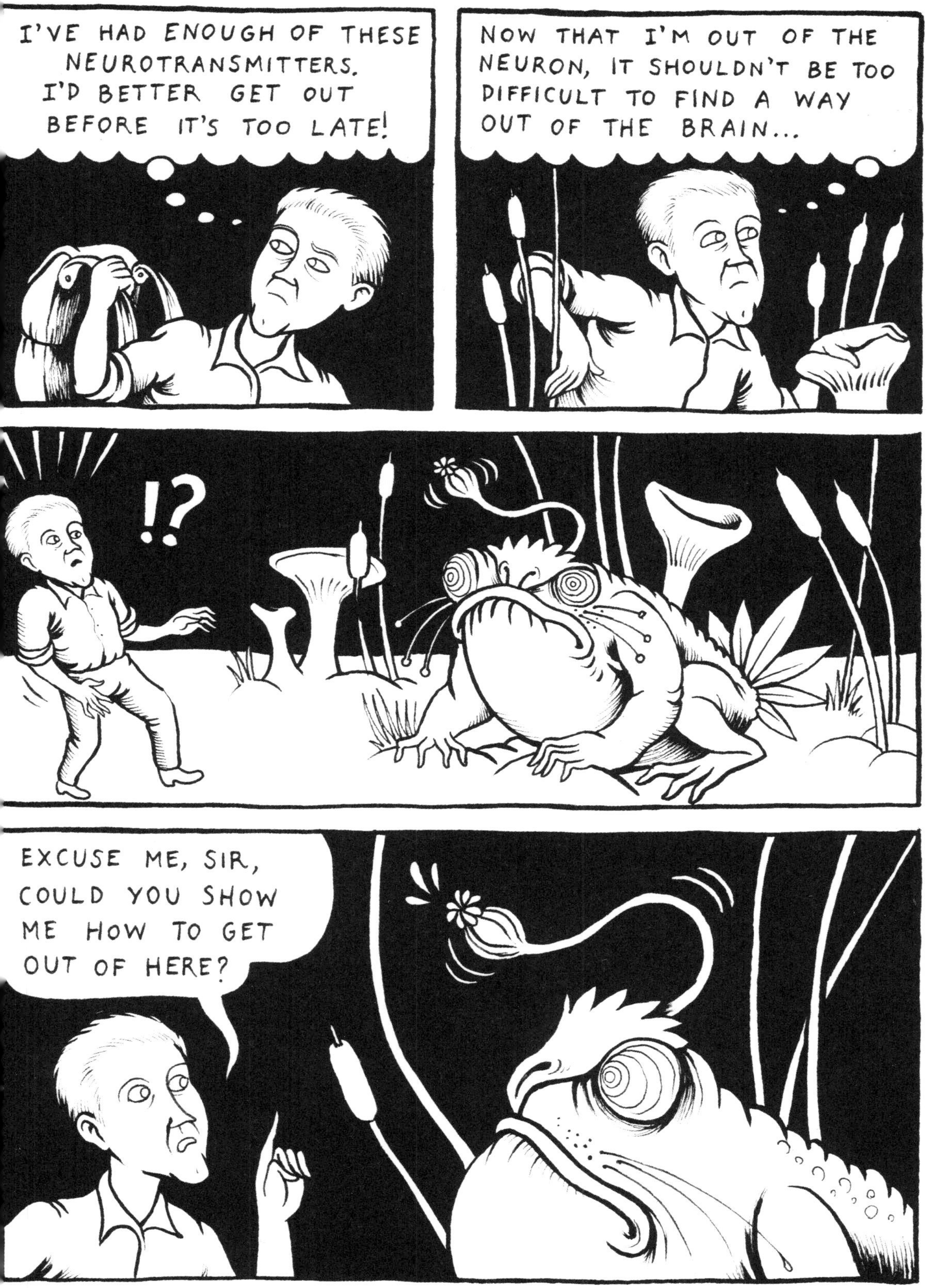

I'VE HAD ENOUGH OF THESE NEUROTRANSMITTERS. I'D BETTER GET OUT BEFORE IT'S TOO LATE!
NOW THAT I'M OUT OF THE NEURON, IT SHOULDN'T BE TOO DIFFICULT TO FIND A WAY OUT OF THE BRAIN...
!?
EXCUSE ME, SIR, COULD YOU SHOW ME HOW TO GET OUT OF HERE?

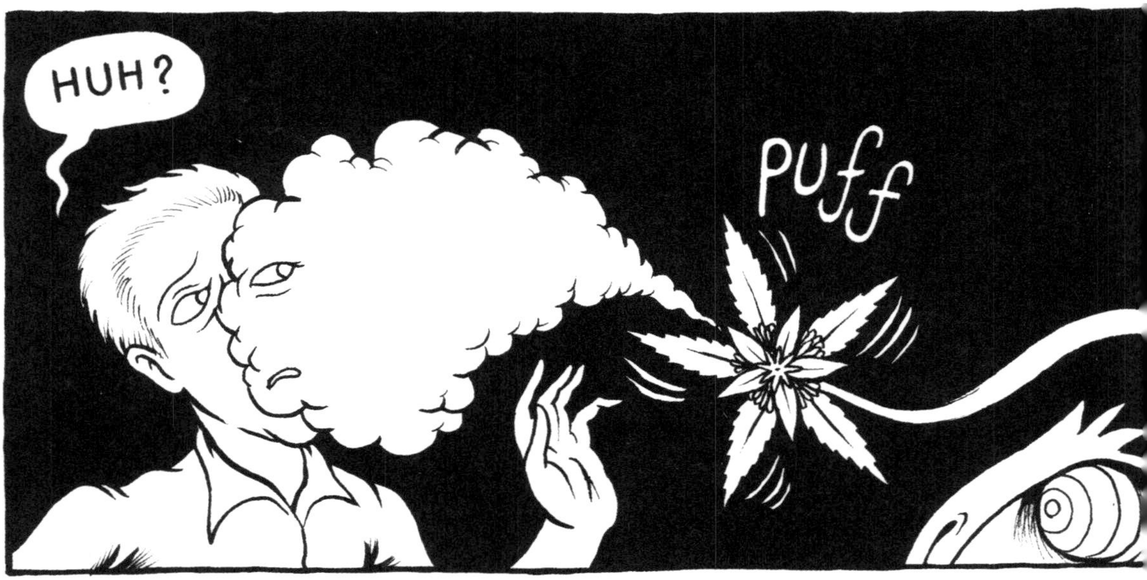
HUH?
PUFF

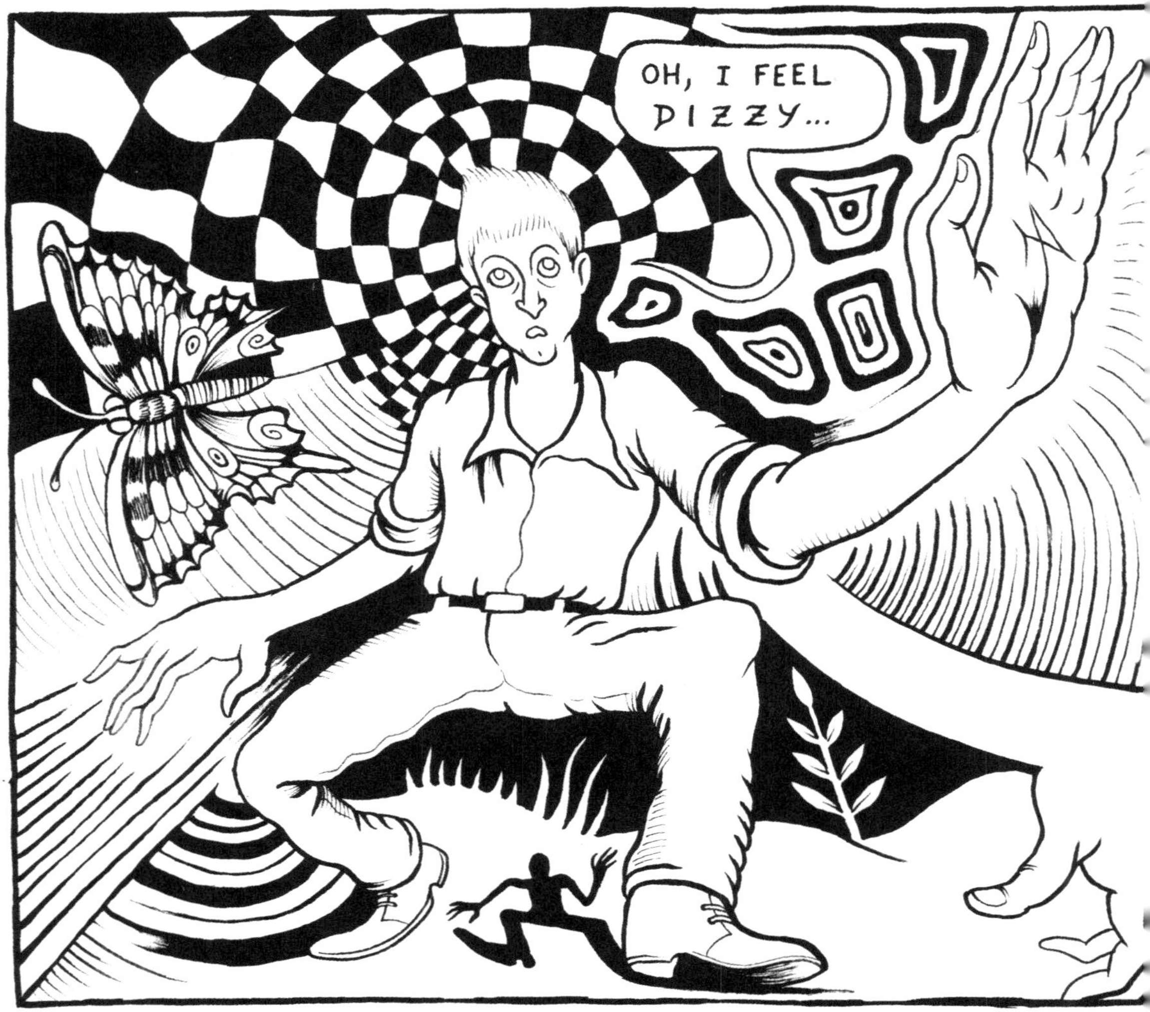
OH, I FEEL DIZZY...

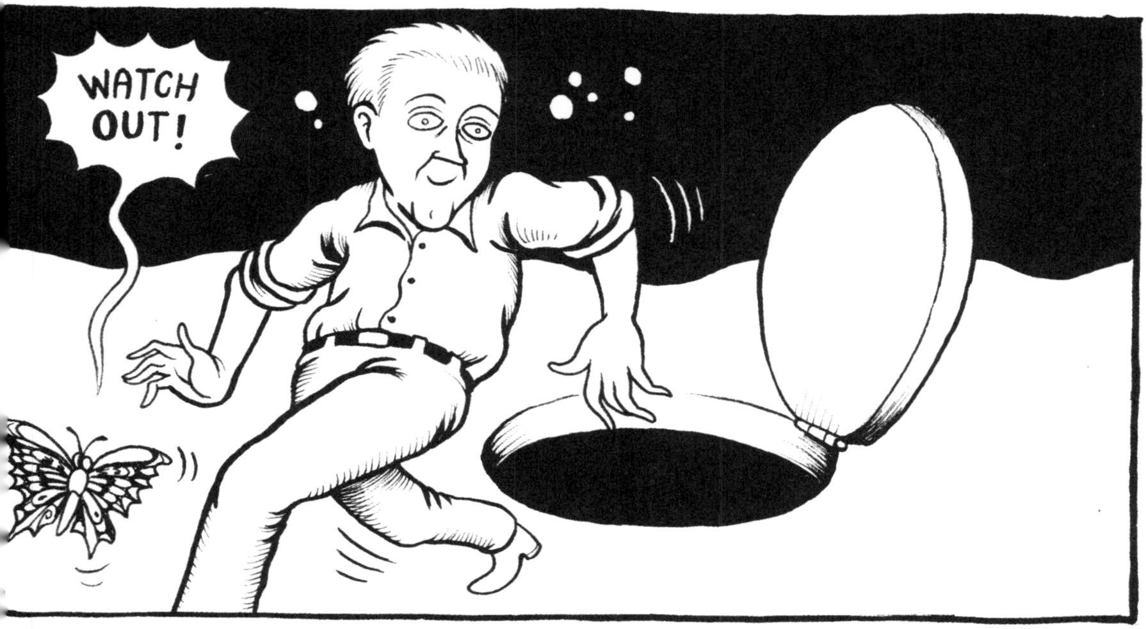
WATCH OUT!

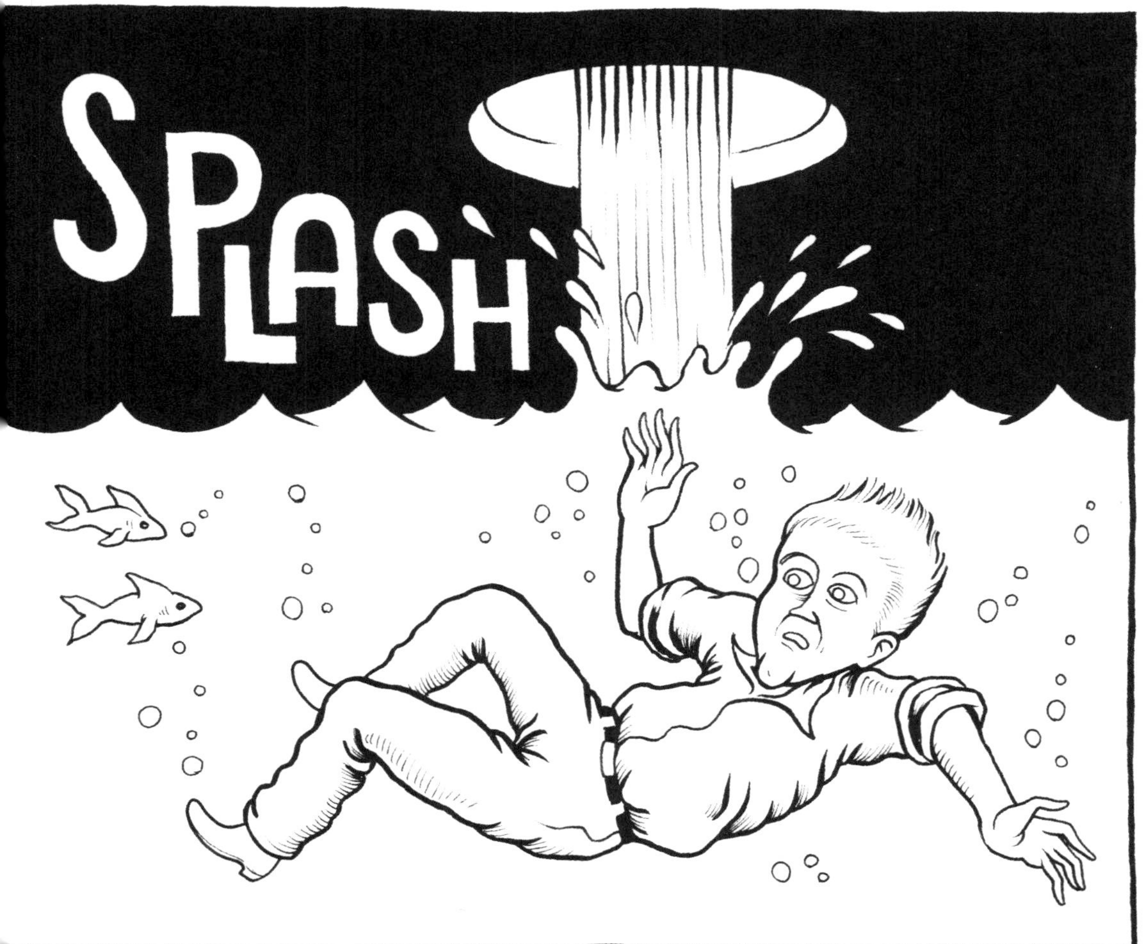
SPLASH

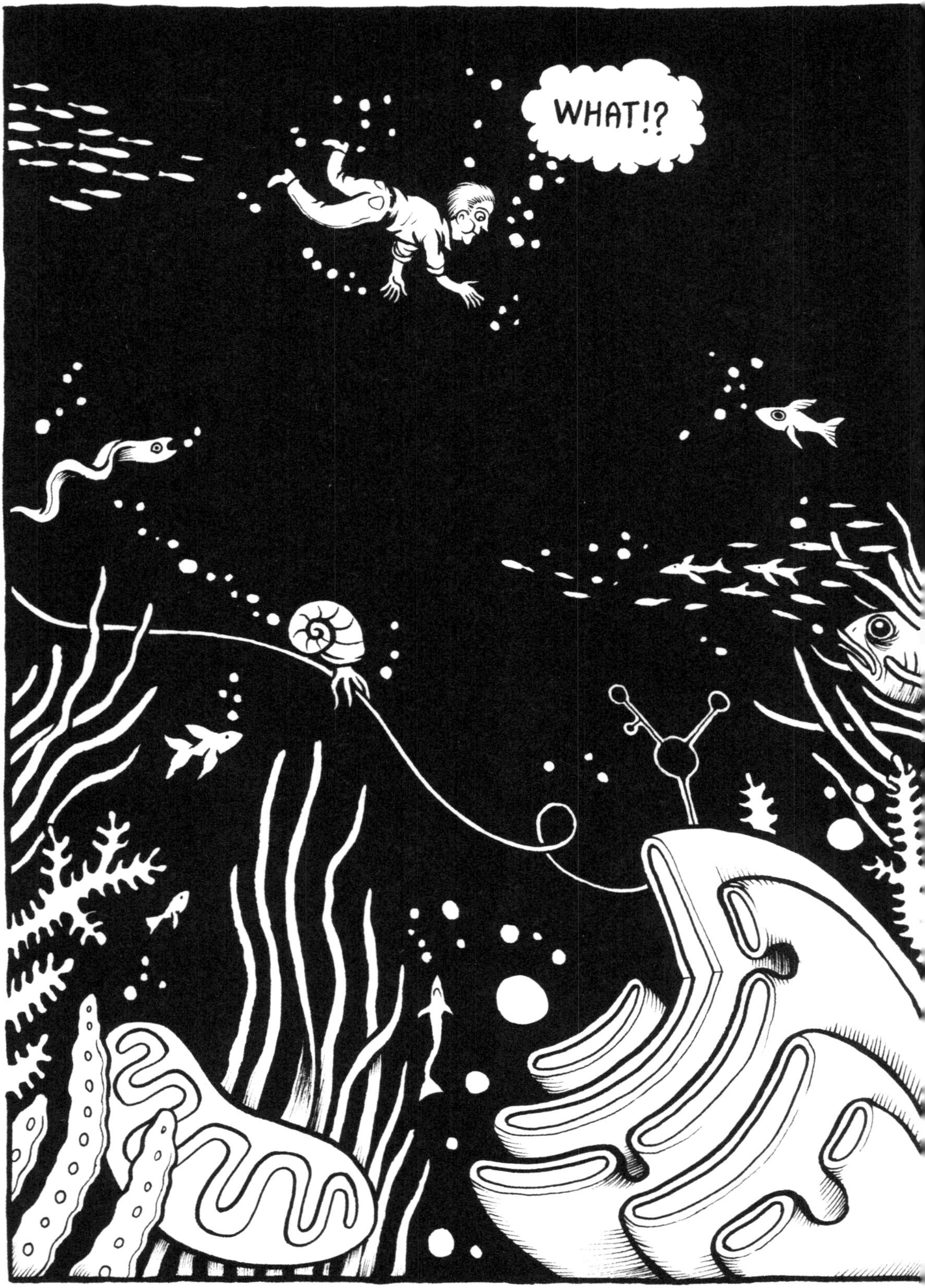
WHAT!?

ELECTROPHYSIOLOGY

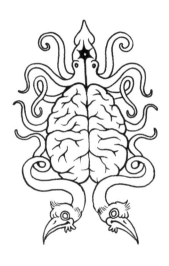

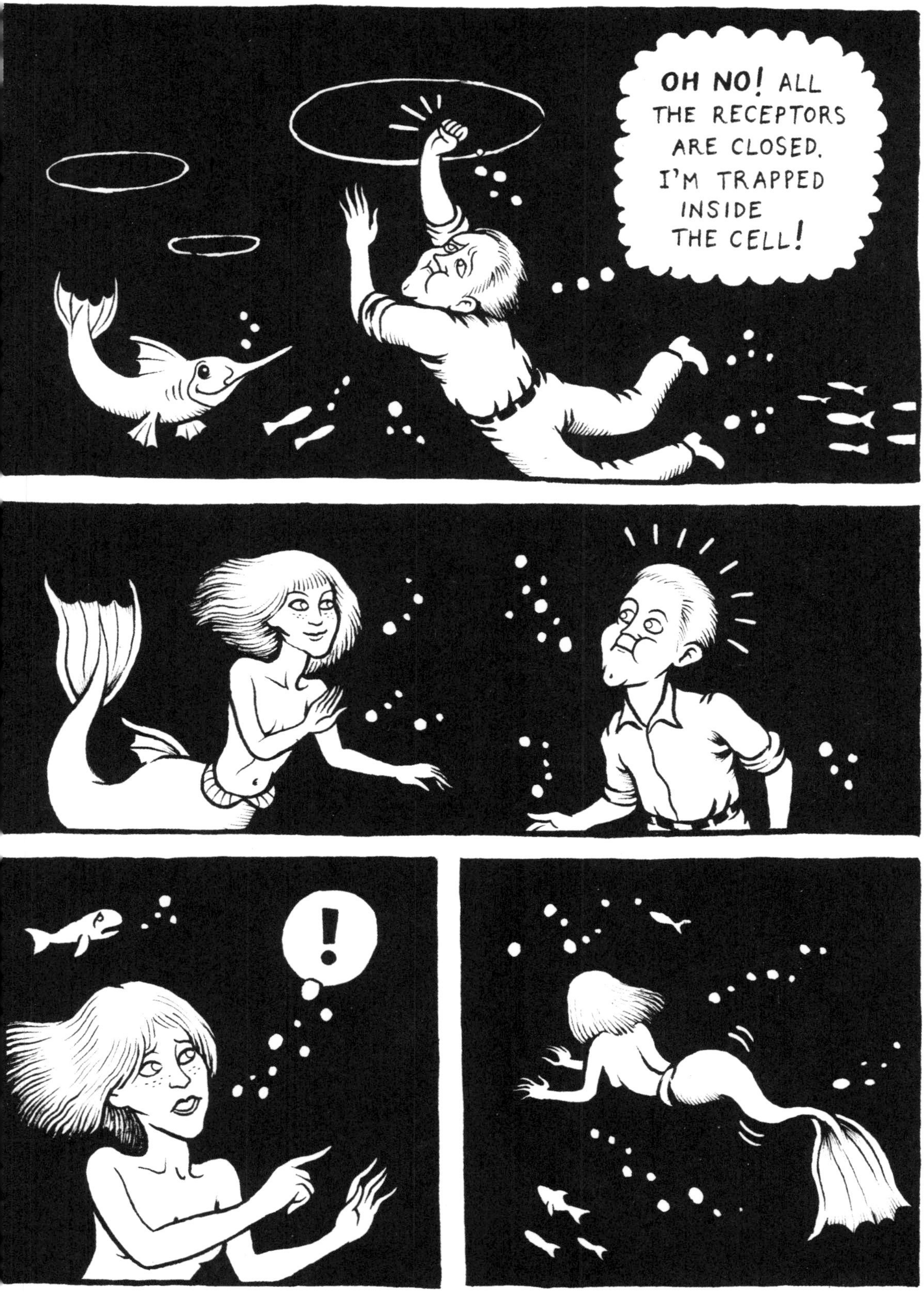
OH NO! ALL THE RECEPTORS ARE CLOSED. I'M TRAPPED INSIDE THE CELL!
!

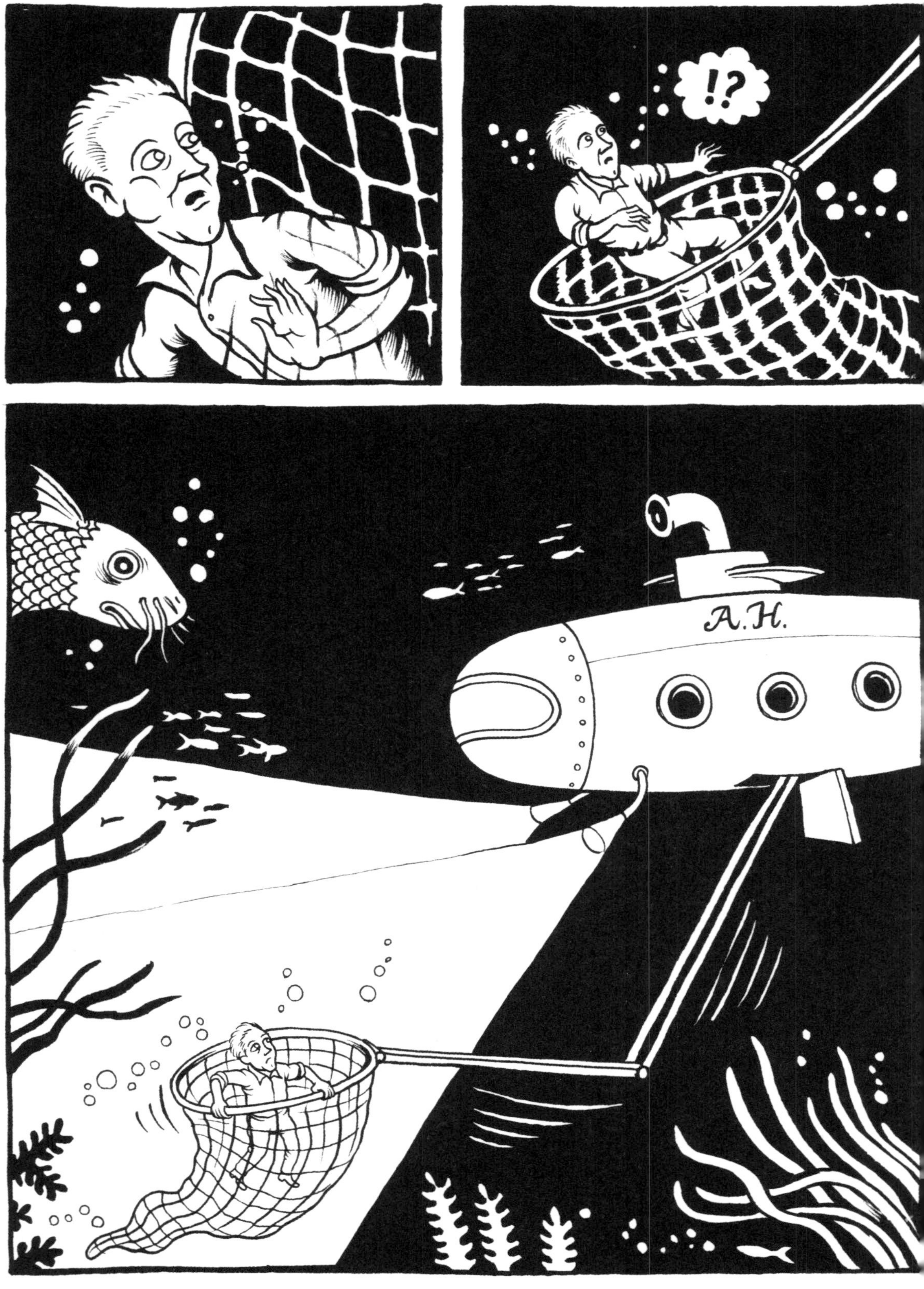
!?
A.H.

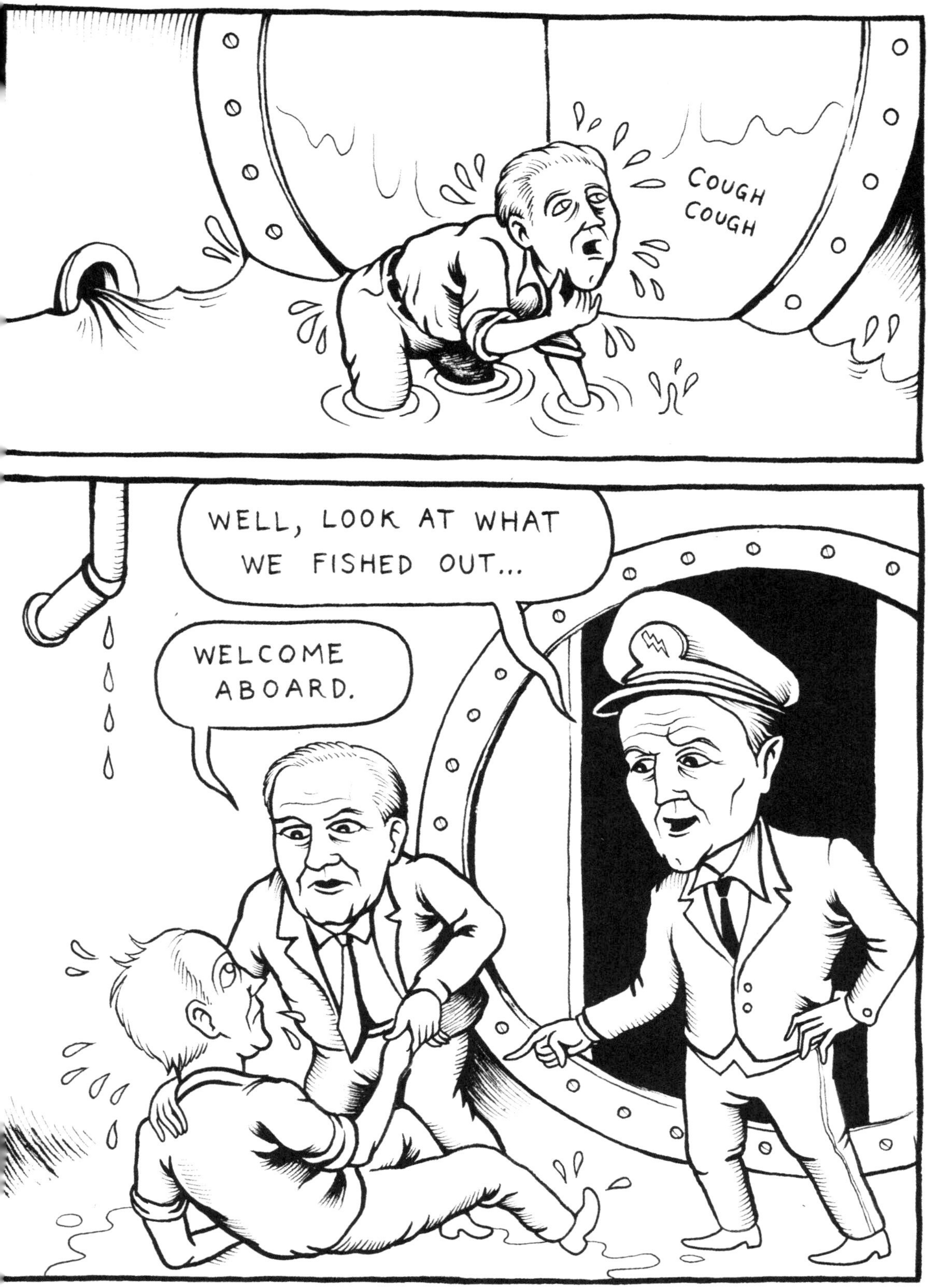
COUGH
COUGH
WELL, LOOK AT WHAT WE FISHED OUT...
WELCOME ABOARD.

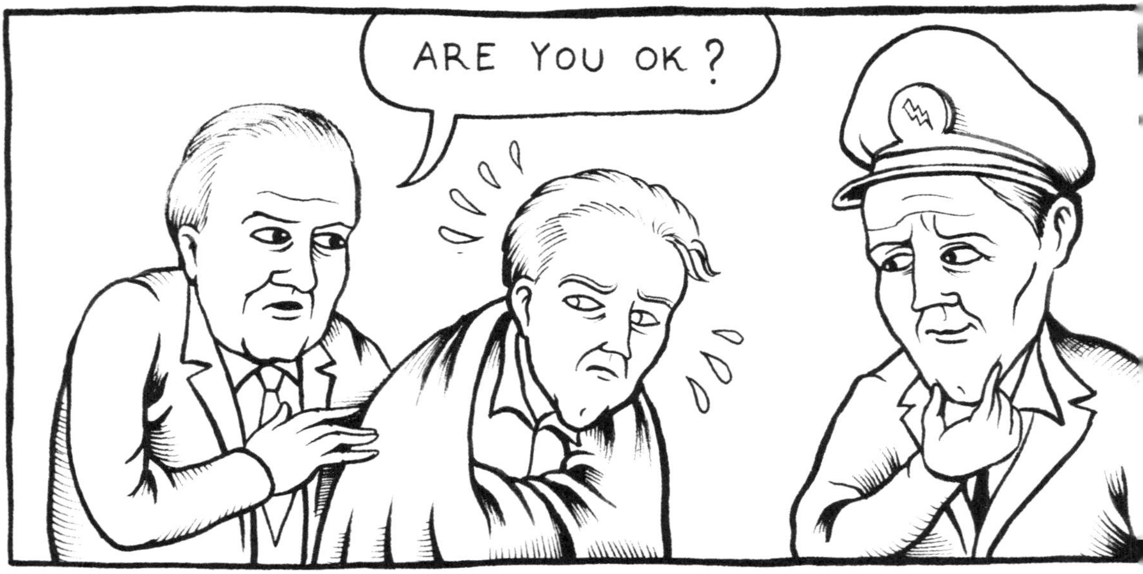
ARE YOU OK?

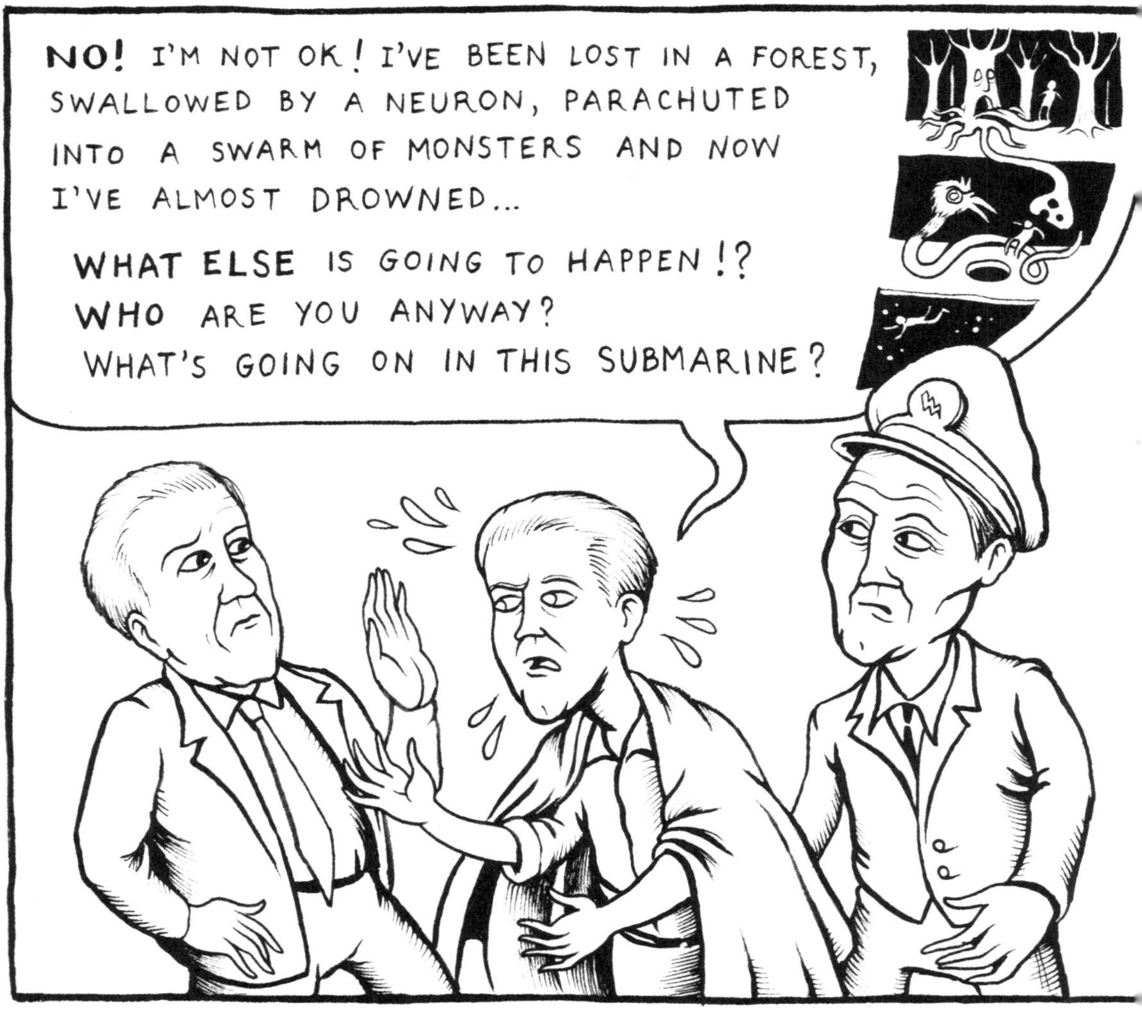
NO! I'M NOT OK! I'VE BEEN LOST IN A FOREST, SWALLOWED BY A NEURON, PARACHUTED INTO A SWARM OF MONSTERS AND NOW I'VE ALMOST DROWNED...
WHAT ELSE IS GOING TO HAPPEN!?
WHO ARE YOU ANYWAY?
WHAT'S GOING ON IN THIS SUBMARINE?

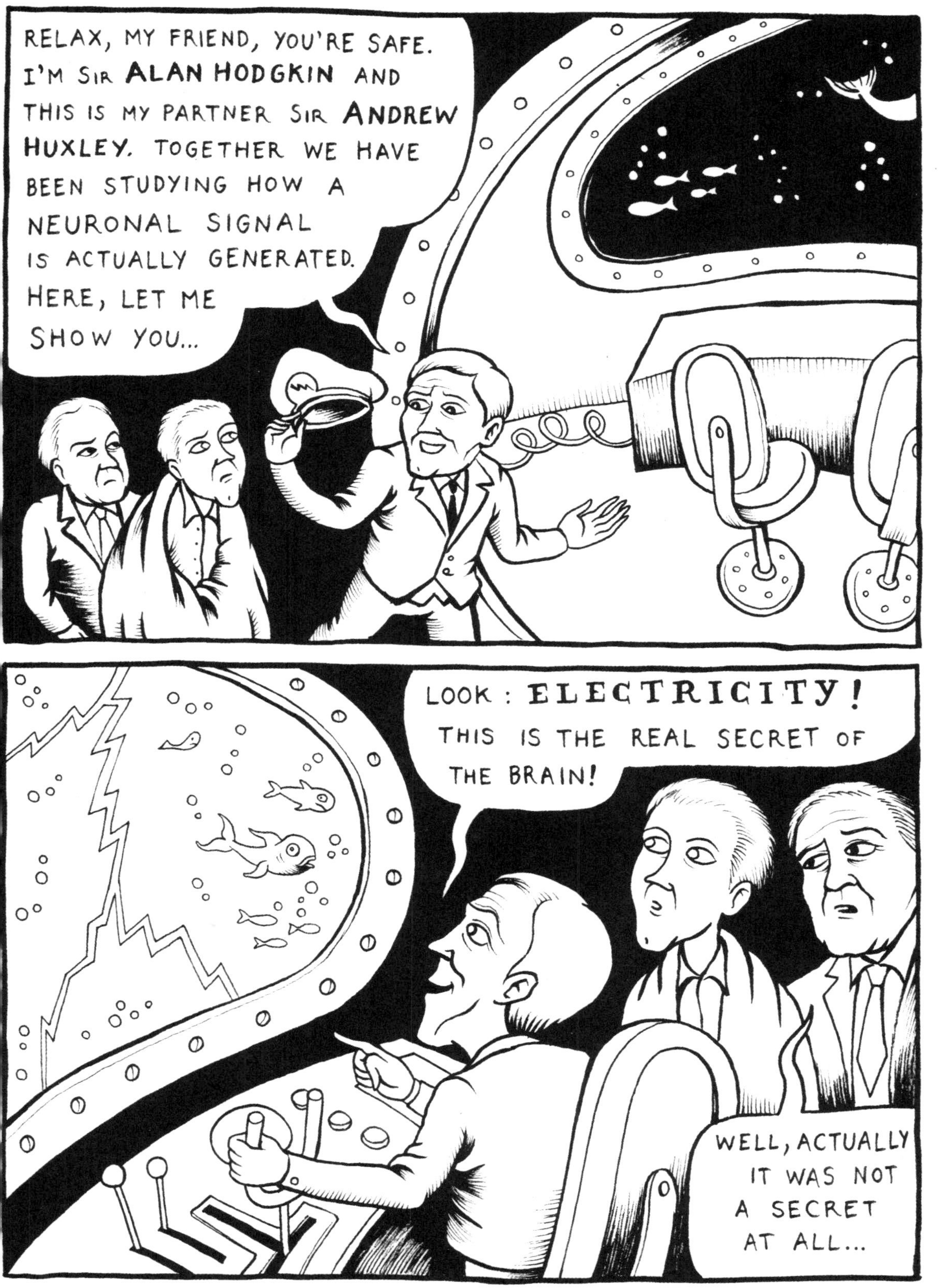
RELAX, MY FRIEND, YOU'RE SAFE. I'M SIR ALAN HODGKIN AND THIS IS MY PARTNER SIR ANDREW HUXLEY. TOGETHER WE HAVE BEEN STUDYING HOW A NEURONAL SIGNAL IS ACTUALLY GENERATED. HERE, LET ME SHOW YOU...
LOOK: ELECTRICITY! THIS IS THE REAL SECRET OF THE BRAIN!
WELL, ACTUALLY IT WAS NOT A SECRET AT ALL...

LONG BEFORE GOLGI STARTED TO LOOK AT NEURONS IN HIS MICROSCOPE, IT WAS KNOWN THAT THE NERVOUS SYSTEM WAS AN ELECTRICAL MACHINE.

IN THE XVI CENTURY, ANOTHER ITALIAN SCIENTIST DISCOVERED THAT MUSCLES CAN BE CONTROLLED BY ELECTRICITY.

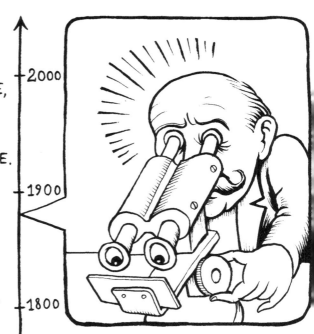

LUIGI GALVANI WAS INTERESTED IN THE EFFECTS OF ELECTRICITY ON THE HUMAN BODY.

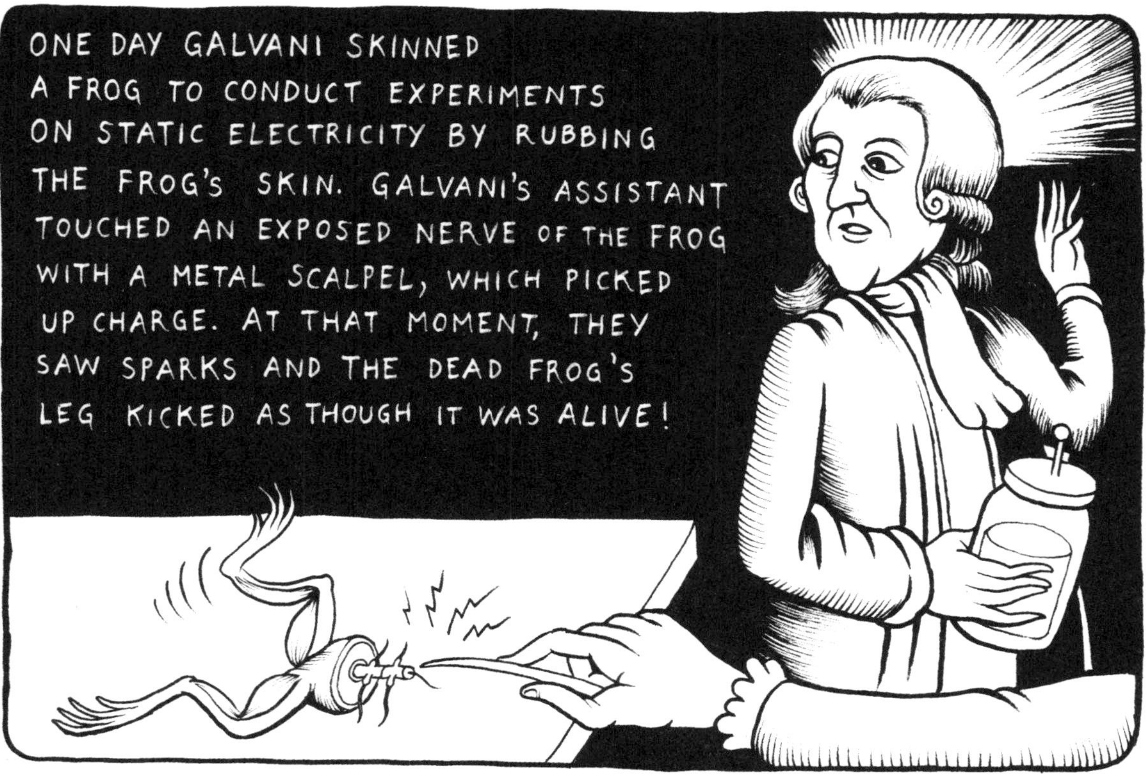
ONE DAY GALVANI SKINNED A FROG TO CONDUCT EXPERIMENTS ON STATIC ELECTRICITY BY RUBBING THE FROG'S SKIN. GALVANI'S ASSISTANT TOUCHED AN EXPOSED NERVE OF THE FROG WITH A METAL SCALPEL, WHICH PICKED UP CHARGE. AT THAT MOMENT, THEY SAW SPARKS AND THE DEAD FROG'S LEG KICKED AS THOUGH IT WAS ALIVE!

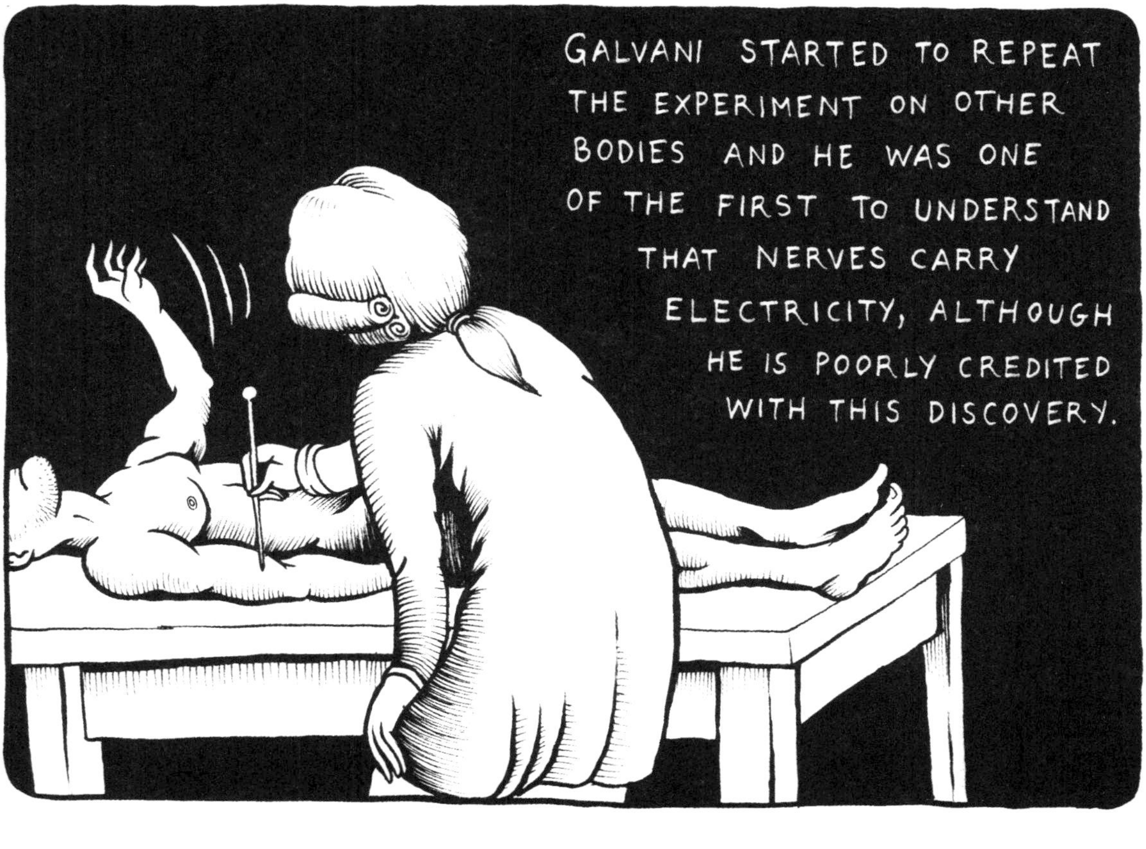
GALVANI STARTED TO REPEAT THE EXPERIMENT ON OTHER BODIES AND HE WAS ONE OF THE FIRST TO UNDERSTAND THAT NERVES CARRY ELECTRICITY, ALTHOUGH HE IS POORLY CREDITED WITH THIS DISCOVERY.

ELECTRICITY ⇝ IS THE FLOW OF **IONS**

(CHARGED PARTICLES) FROM ONE REGION TO ANOTHER. YOU SEE, IONS OF THE SAME CHARGE REALLY DON'T LIKE EACH OTHER, SO IF MANY OF THEM ARE CONCENTRATED INSIDE A MEMBRANE THEY'LL TEND TO ESCAPE THROUGH PERMEABLE HOLES, PRODUCING **ELECTRIC CURRENTS.**

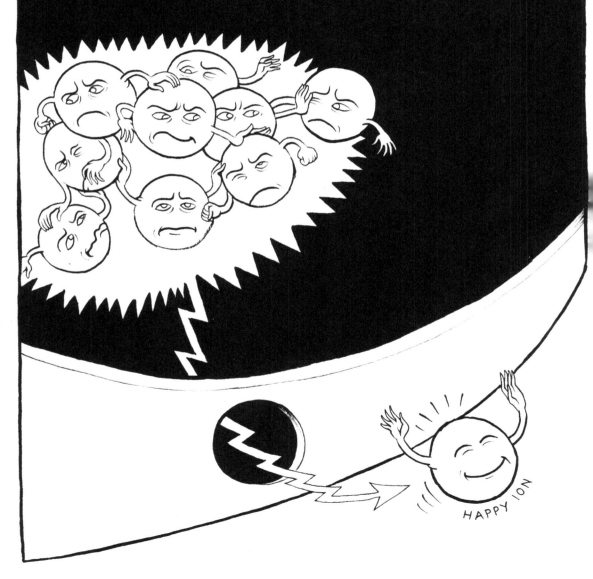

HAPPY ION

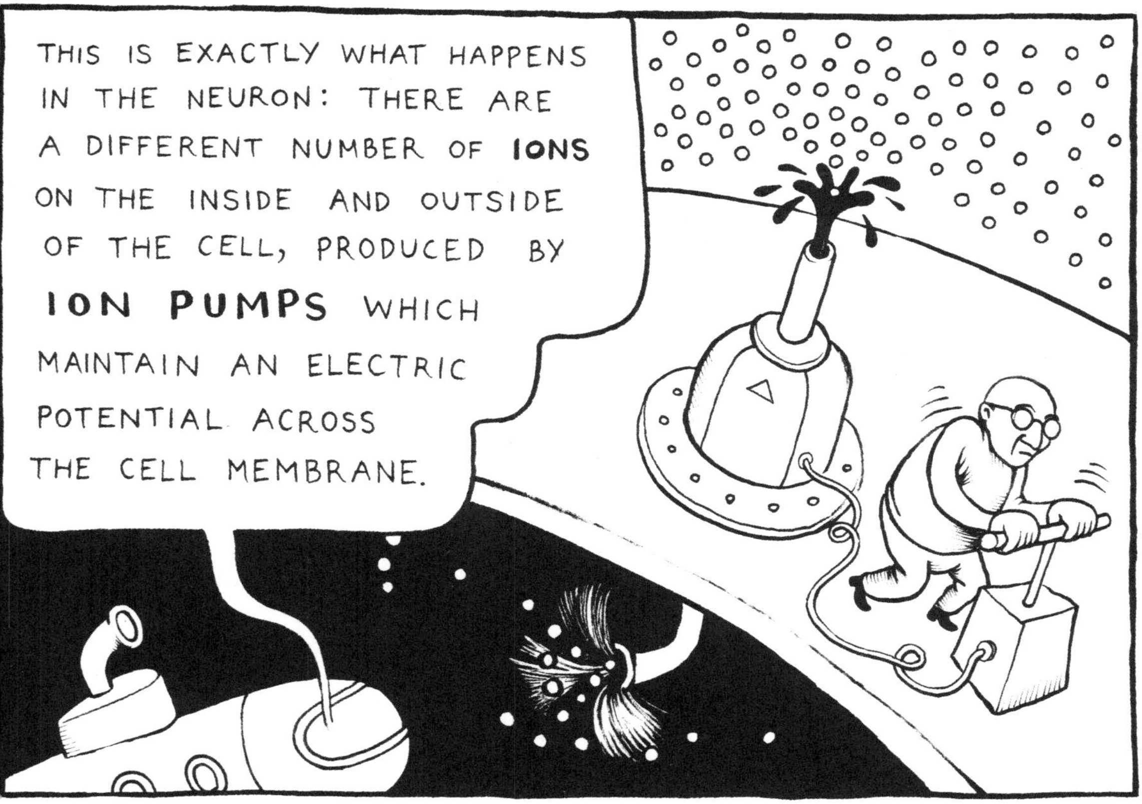
THIS IS EXACTLY WHAT HAPPENS IN THE NEURON: THERE ARE A DIFFERENT NUMBER OF **IONS** ON THE INSIDE AND OUTSIDE OF THE CELL, PRODUCED BY **ION PUMPS** WHICH MAINTAIN AN ELECTRIC POTENTIAL ACROSS THE CELL MEMBRANE.

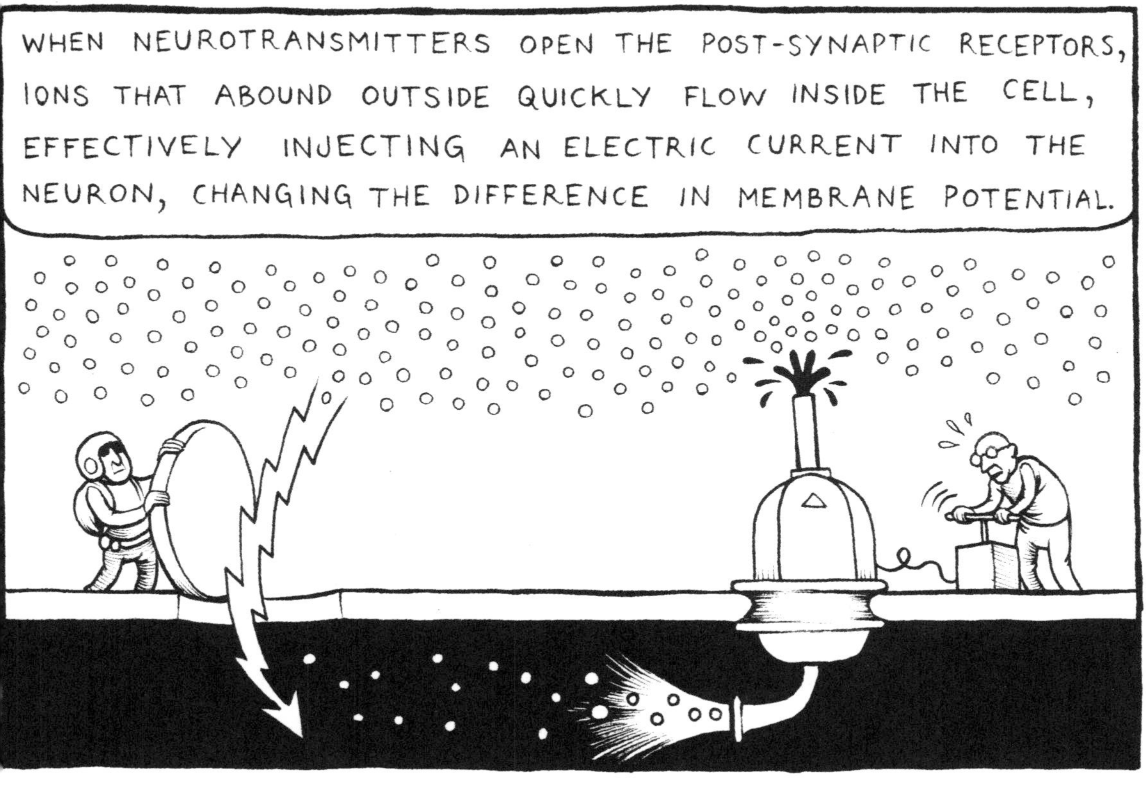
WHEN NEUROTRANSMITTERS OPEN THE POST-SYNAPTIC RECEPTORS, IONS THAT ABOUND OUTSIDE QUICKLY FLOW INSIDE THE CELL, EFFECTIVELY INJECTING AN ELECTRIC CURRENT INTO THE NEURON, CHANGING THE DIFFERENCE IN MEMBRANE POTENTIAL.

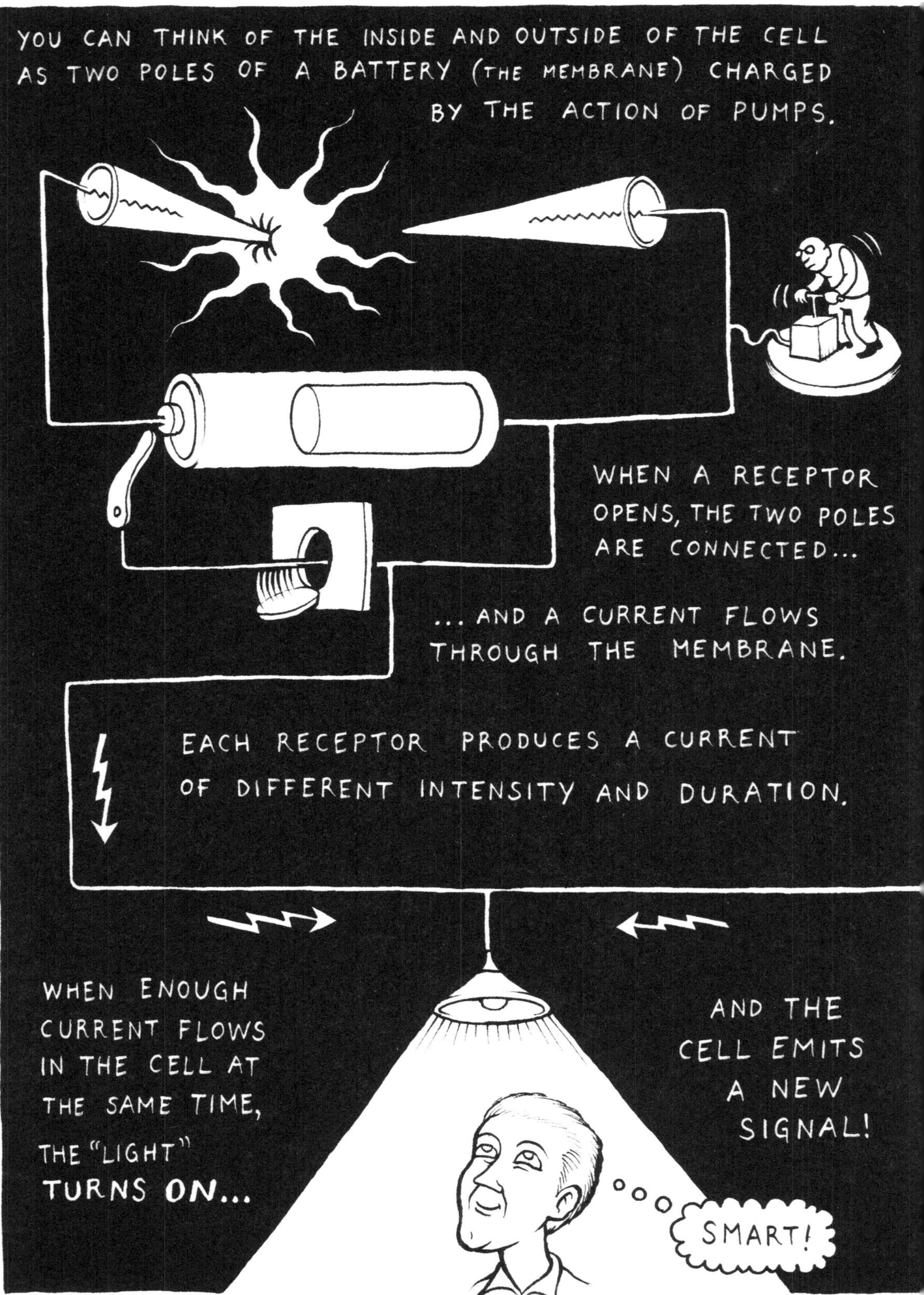

YOU CAN THINK OF THE INSIDE AND OUTSIDE OF THE CELL AS TWO POLES OF A BATTERY (THE MEMBRANE) CHARGED BY THE ACTION OF PUMPS.

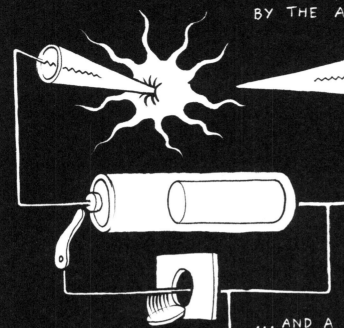

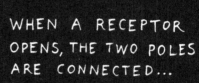

WHEN A RECEPTOR OPENS, THE TWO POLES ARE CONNECTED...

... AND A CURRENT FLOWS THROUGH THE MEMBRANE.

EACH RECEPTOR PRODUCES A CURRENT OF DIFFERENT INTENSITY AND DURATION.

WHEN ENOUGH CURRENT FLOWS IN THE CELL AT THE SAME TIME, THE "LIGHT" TURNS **ON**...

AND THE CELL EMITS A NEW SIGNAL!

SMART!

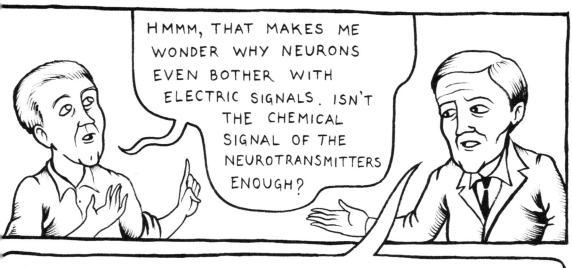

HMMM, THAT MAKES ME WONDER WHY NEURONS EVEN BOTHER WITH ELECTRIC SIGNALS. ISN'T THE CHEMICAL SIGNAL OF THE NEUROTRANSMITTERS ENOUGH?

NO, NO, NO, THE BRAIN NEEDS BOTH! FIRST OF ALL BECAUSE THE ELECTRIC SIGNAL IS FASTER AND IT CAN QUICKLY TRAVEL OVER LONG DISTANCES IN THE BODY, WHICH CAN MEAN THE DIFFERENCE BETWEEN LIFE AND DEATH.

BUT ALSO BECAUSE TRANSLATING ALL THE DIFFERENT CHEMICAL SIGNALS INTO ELECTRIC CURRENTS ENABLES THE NEURON TO COMBINE THEM TOGETHER AND PERFORM COMPUTATIONS INSIDE THE CELL.

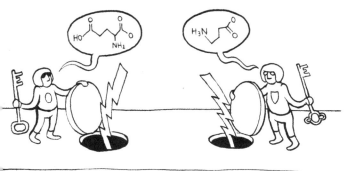

BUT HOW ARE ALL THESE CURRENTS ACTUALLY COMBINED?

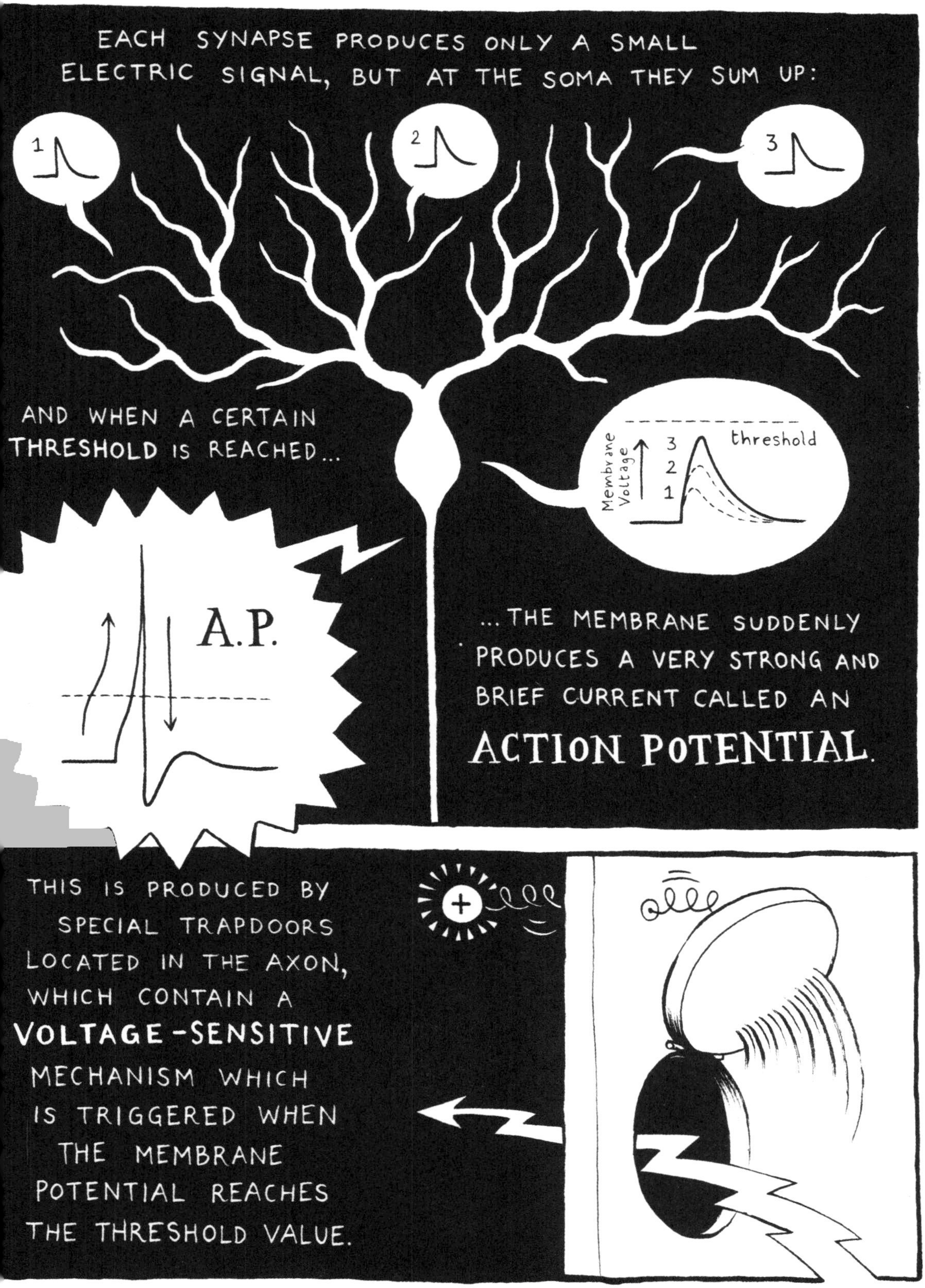

EACH SYNAPSE PRODUCES ONLY A SMALL ELECTRIC SIGNAL, BUT AT THE SOMA THEY SUM UP:
1
2
3
AND WHEN A CERTAIN THRESHOLD IS REACHED...
threshold
Membrane Voltage
3
2
1
A.P.
... THE MEMBRANE SUDDENLY PRODUCES A VERY STRONG AND BRIEF CURRENT CALLED AN ACTION POTENTIAL.
THIS IS PRODUCED BY SPECIAL TRAPDOORS LOCATED IN THE AXON, WHICH CONTAIN A VOLTAGE-SENSITIVE MECHANISM WHICH IS TRIGGERED WHEN THE MEMBRANE POTENTIAL REACHES THE THRESHOLD VALUE.
+

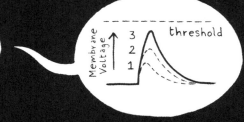

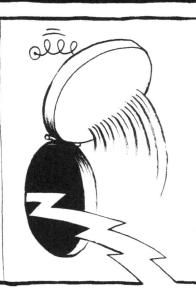

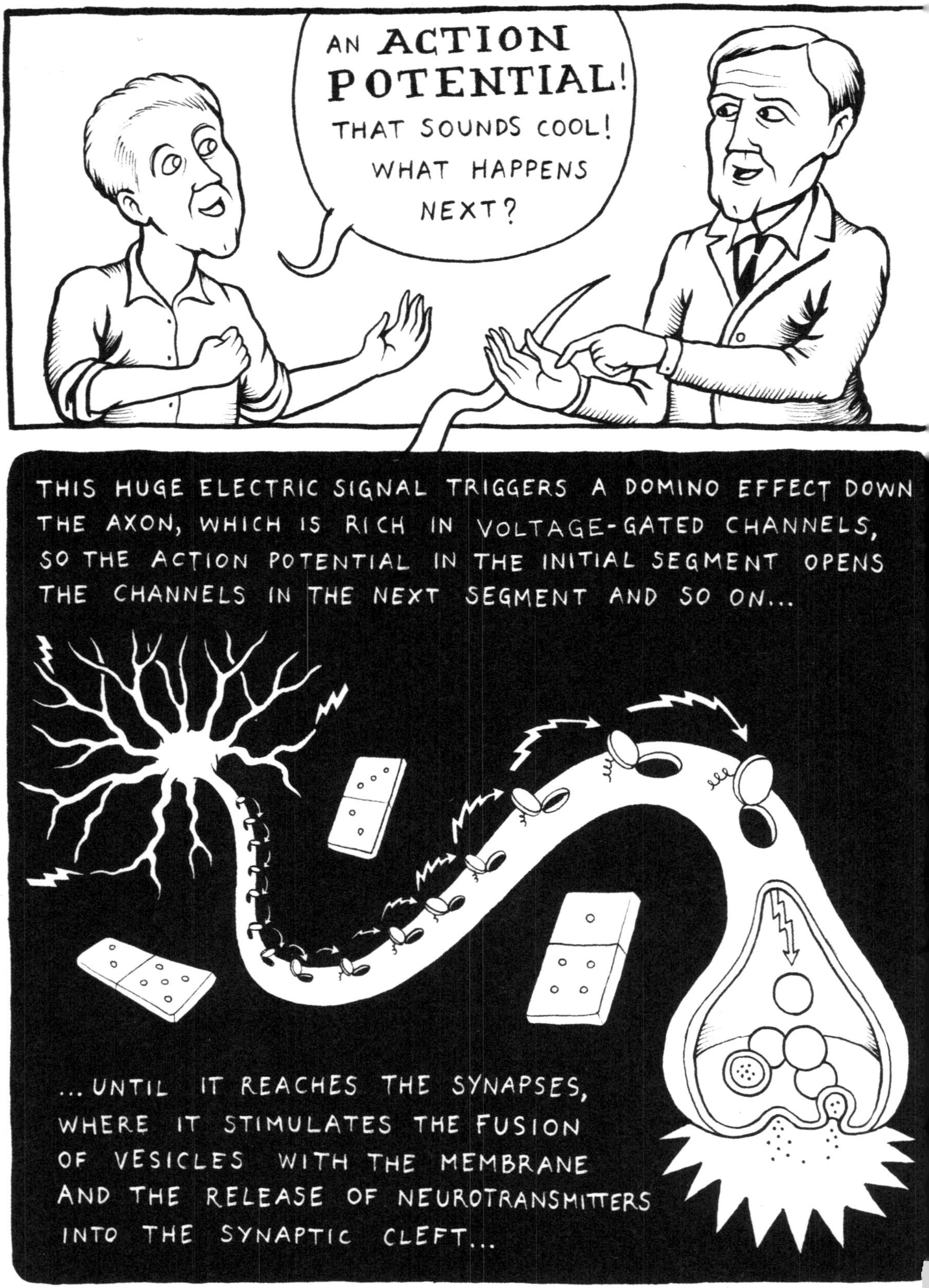
AN ACTION POTENTIAL! THAT SOUNDS COOL! WHAT HAPPENS NEXT?
THIS HUGE ELECTRIC SIGNAL TRIGGERS A DOMINO EFFECT DOWN THE AXON, WHICH IS RICH IN VOLTAGE-GATED CHANNELS, SO THE ACTION POTENTIAL IN THE INITIAL SEGMENT OPENS THE CHANNELS IN THE NEXT SEGMENT AND SO ON...
...UNTIL IT REACHES THE SYNAPSES, WHERE IT STIMULATES THE FUSION OF VESICLES WITH THE MEMBRANE AND THE RELEASE OF NEUROTRANSMITTERS INTO THE SYNAPTIC CLEFT...

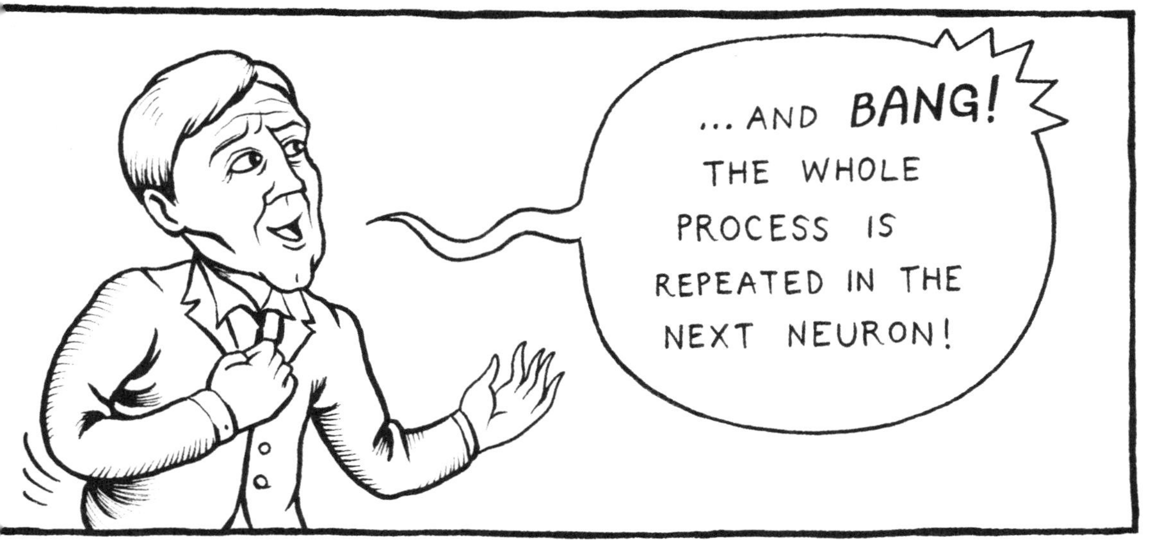
...AND BANG! THE WHOLE PROCESS IS REPEATED IN THE NEXT NEURON!
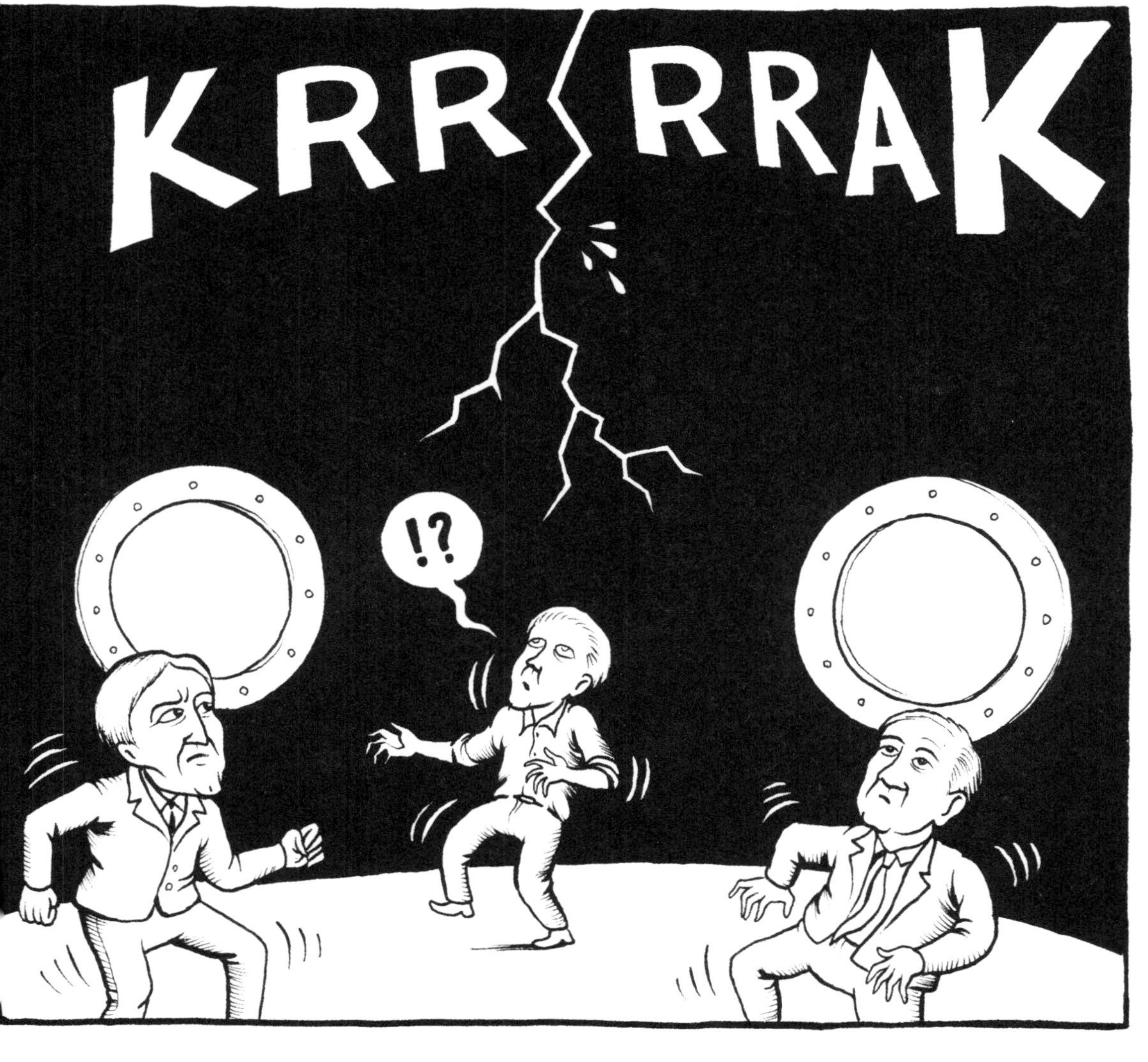
KRRRRAK
!?

HEY, WHAT THE HELL WAS THAT?

THE **KRAKEN!**

WHO?

WELL, YOU SEE...
IT WAS ALMOST IMPOSSIBLE TO RECORD THESE ELECTRIC CURRENTS FROM TINY STRUCTURES LIKE NEURONS.

SO, FOR OUR EXPERIMENTS WE HAD TO USE SQUID, WHICH HAVE GIANT AXONS OF 1mm IN DIAMETER...

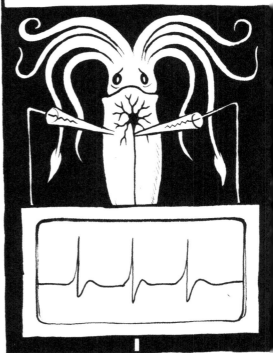

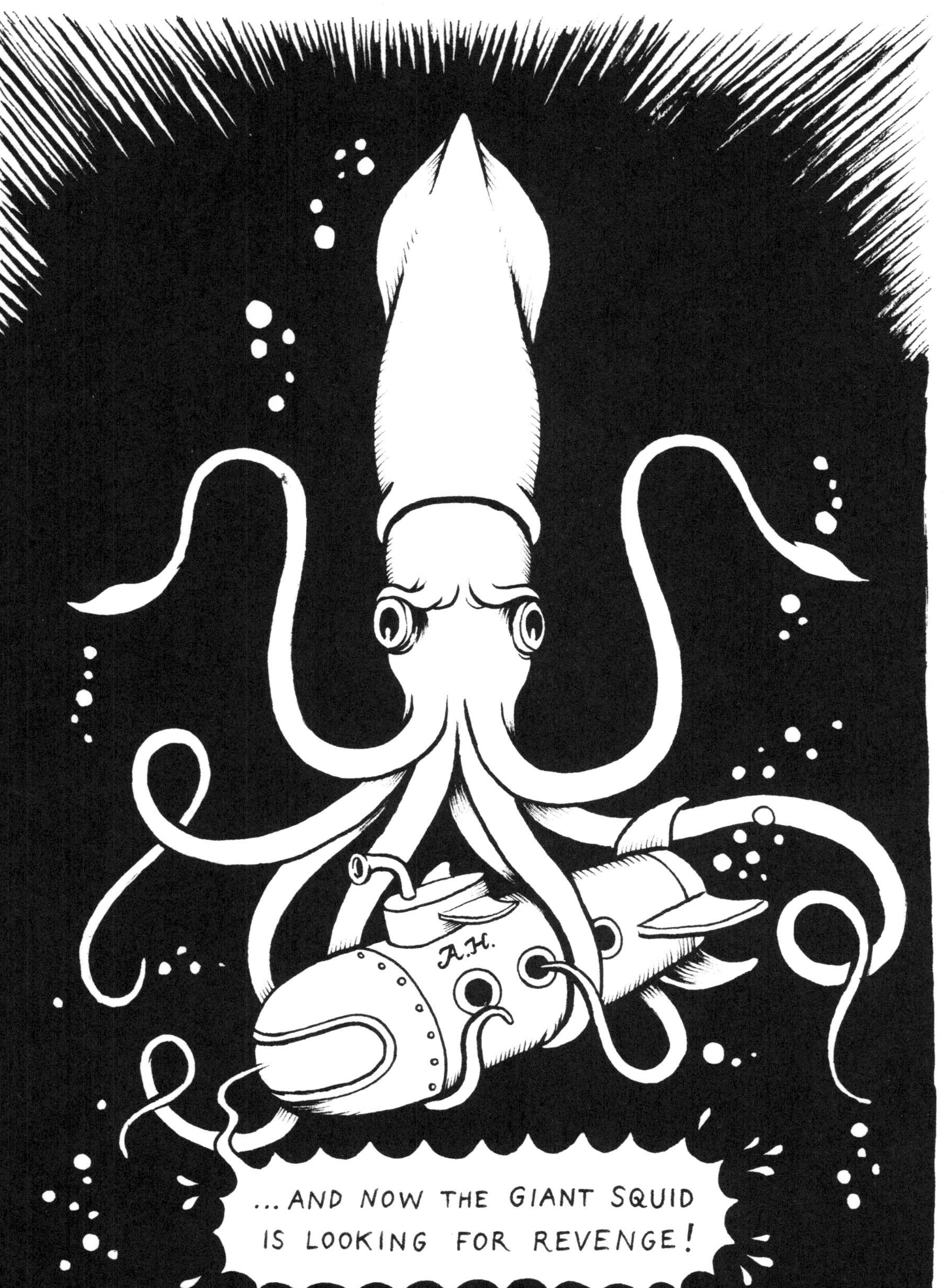
A.H.
...AND NOW THE GIANT SQUID IS LOOKING FOR REVENGE!

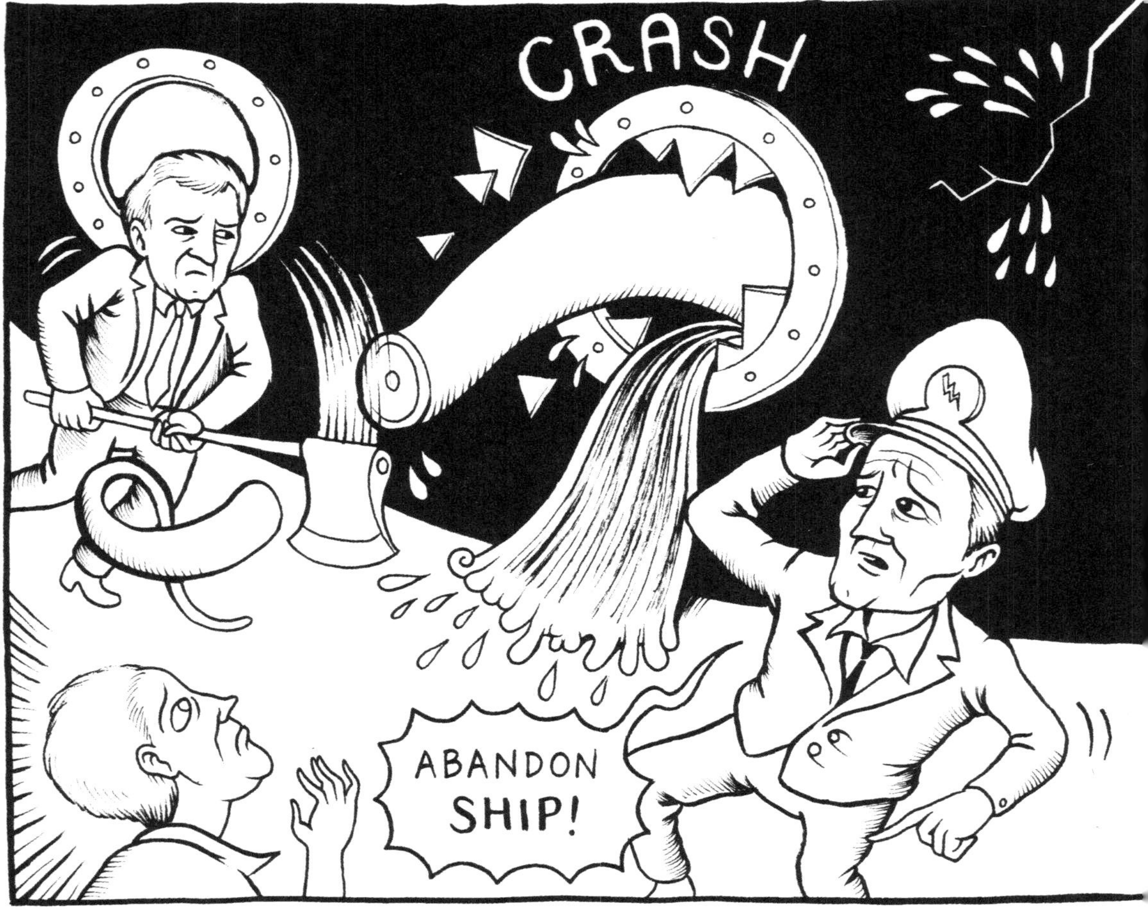
CRASH
ABANDON SHIP!

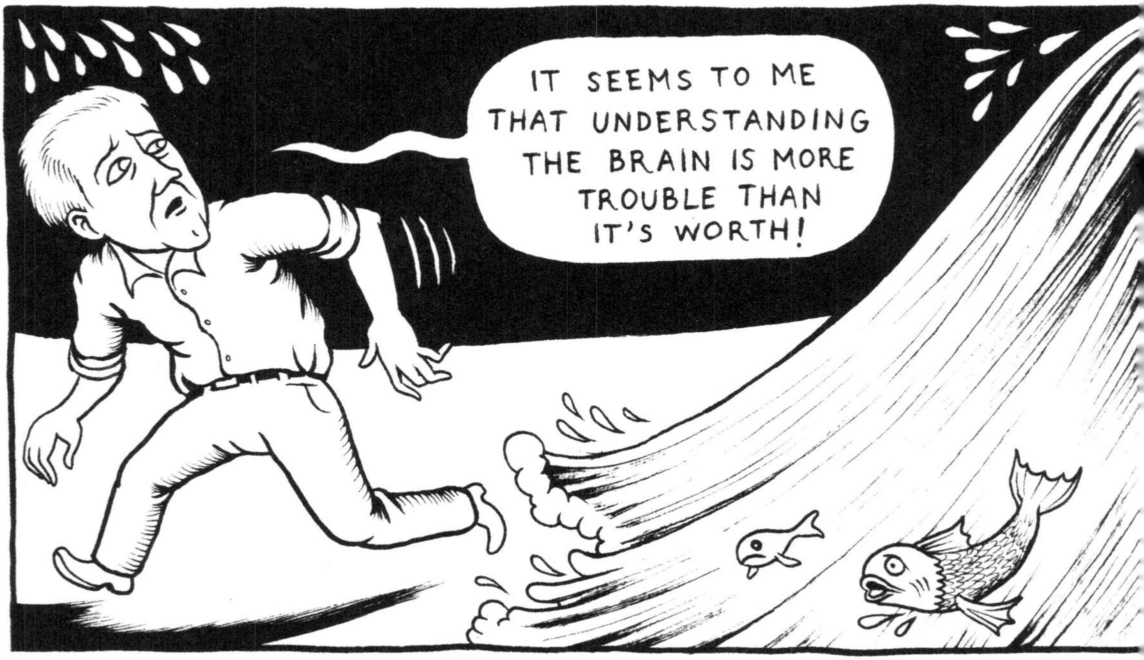
IT SEEMS TO ME THAT UNDERSTANDING THE BRAIN IS MORE TROUBLE THAN IT'S WORTH!

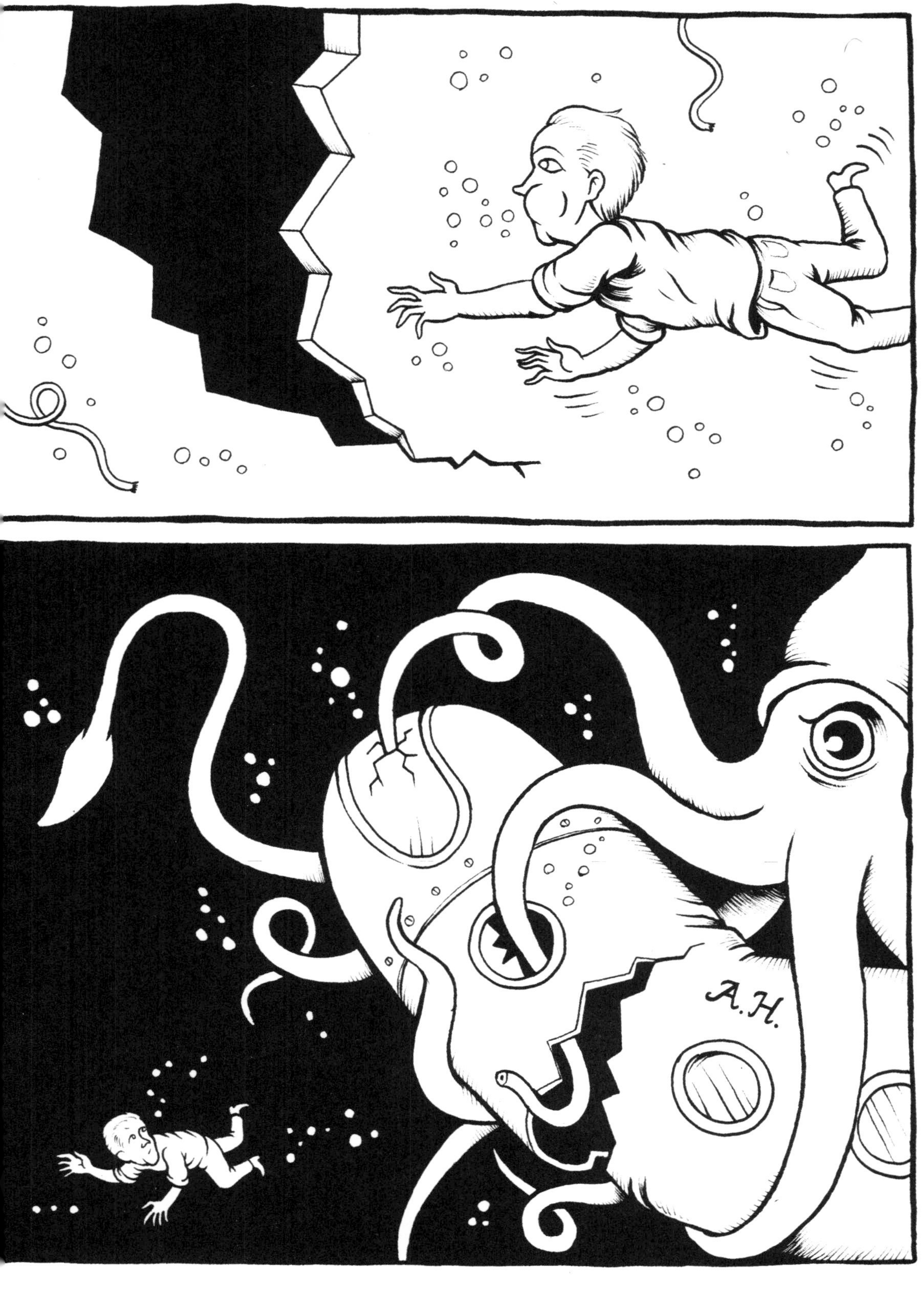
A.H.

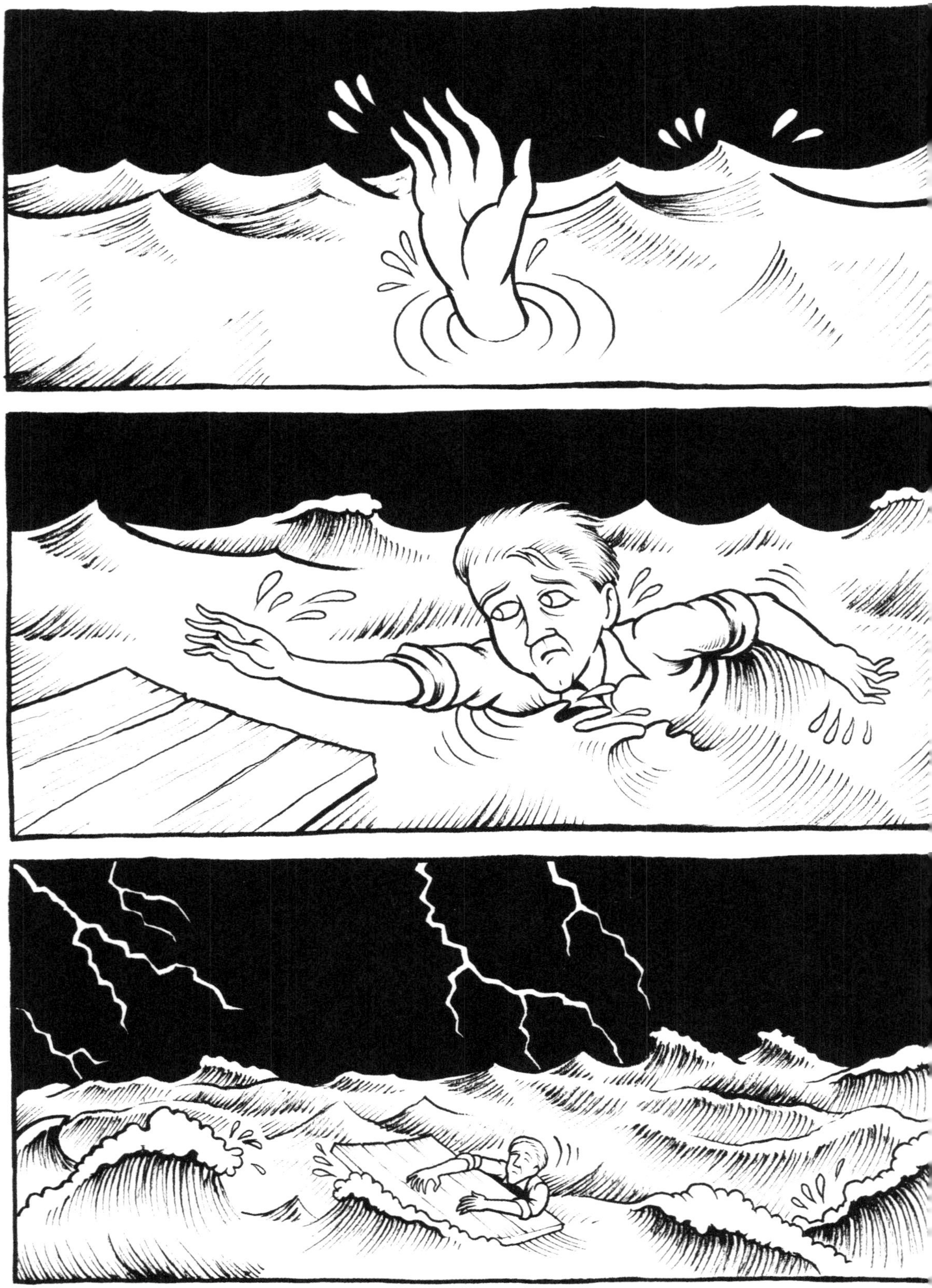

PLASTICITY

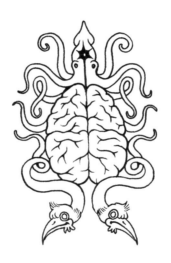

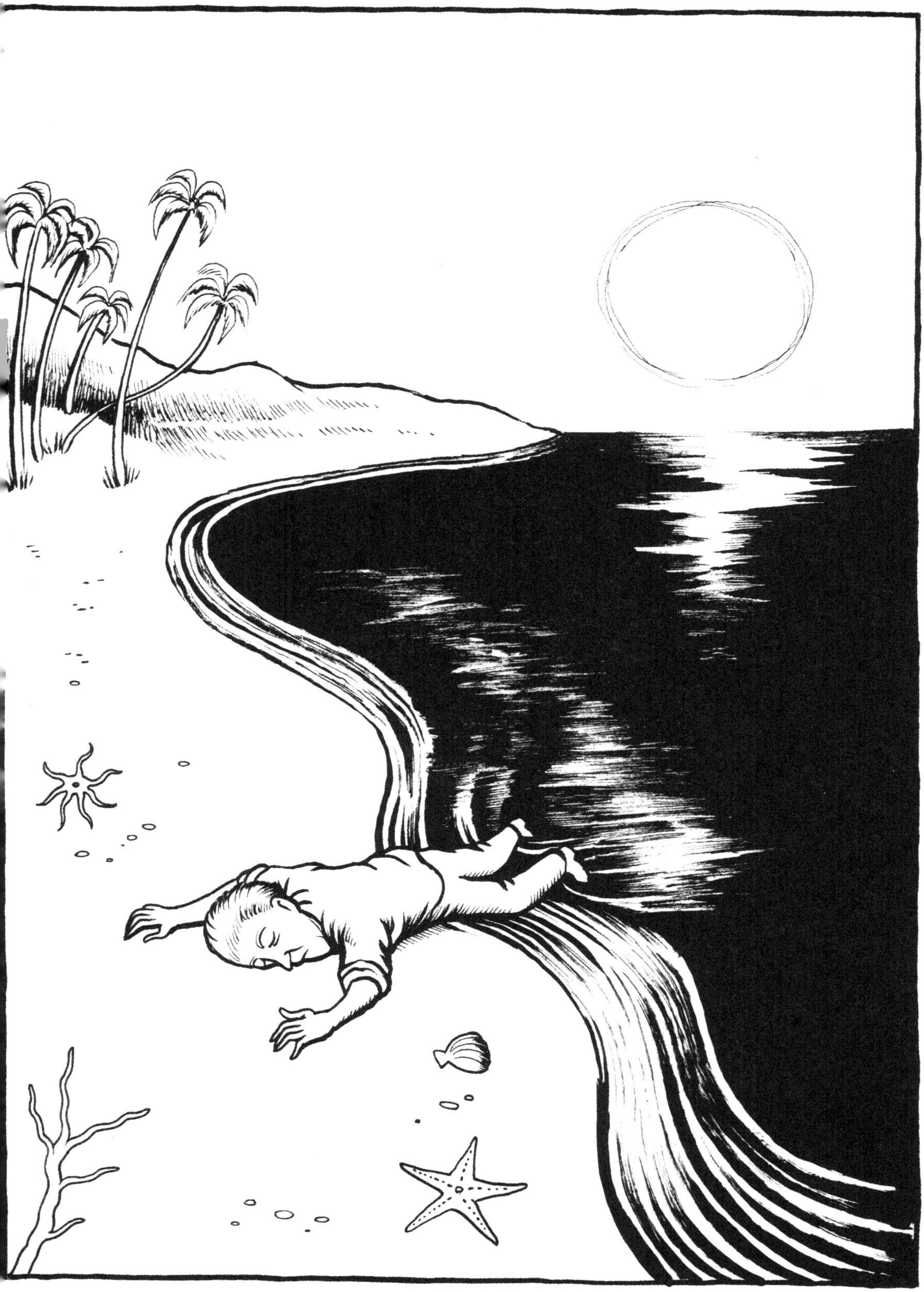

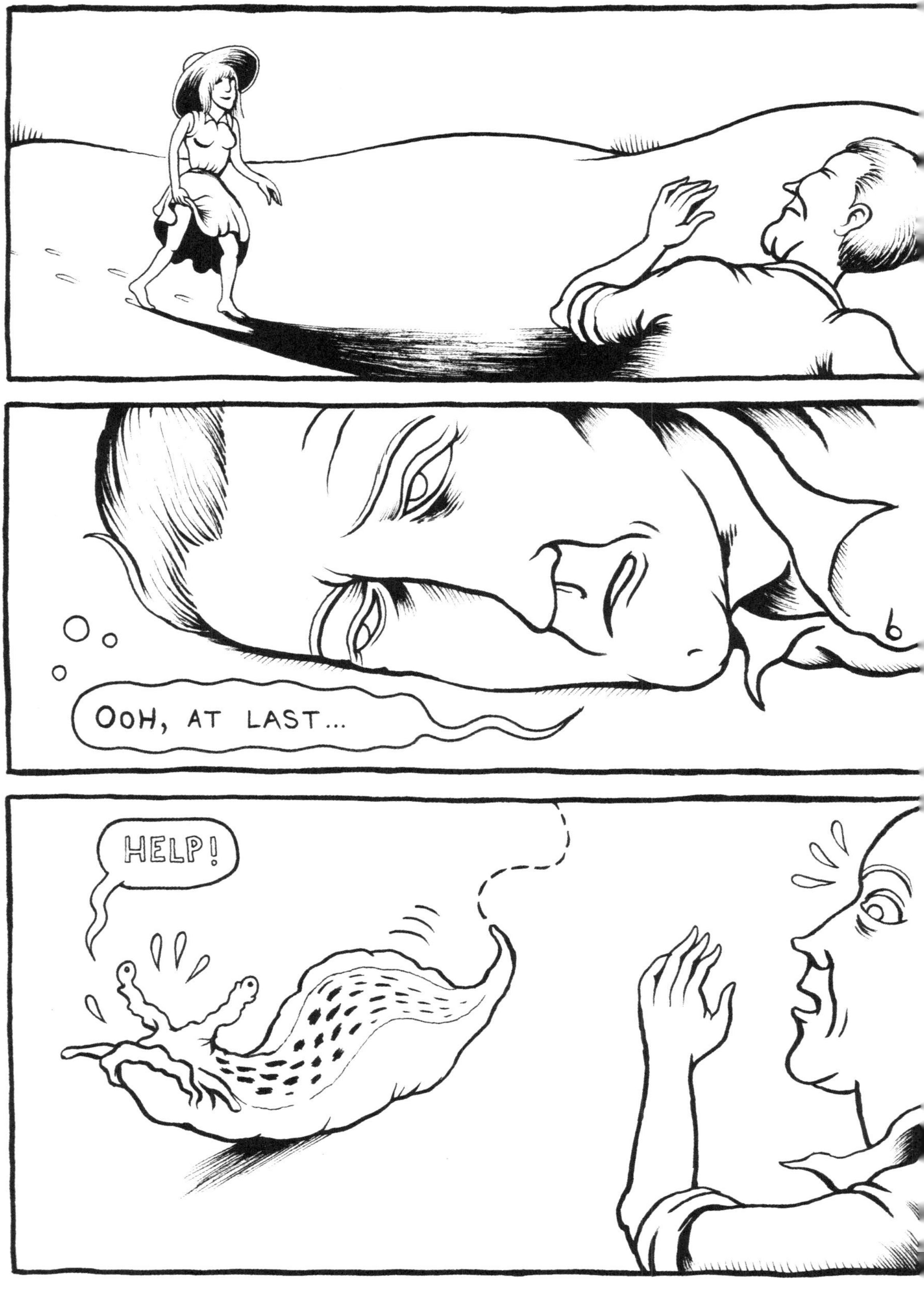
OOH, AT LAST...
HELP!

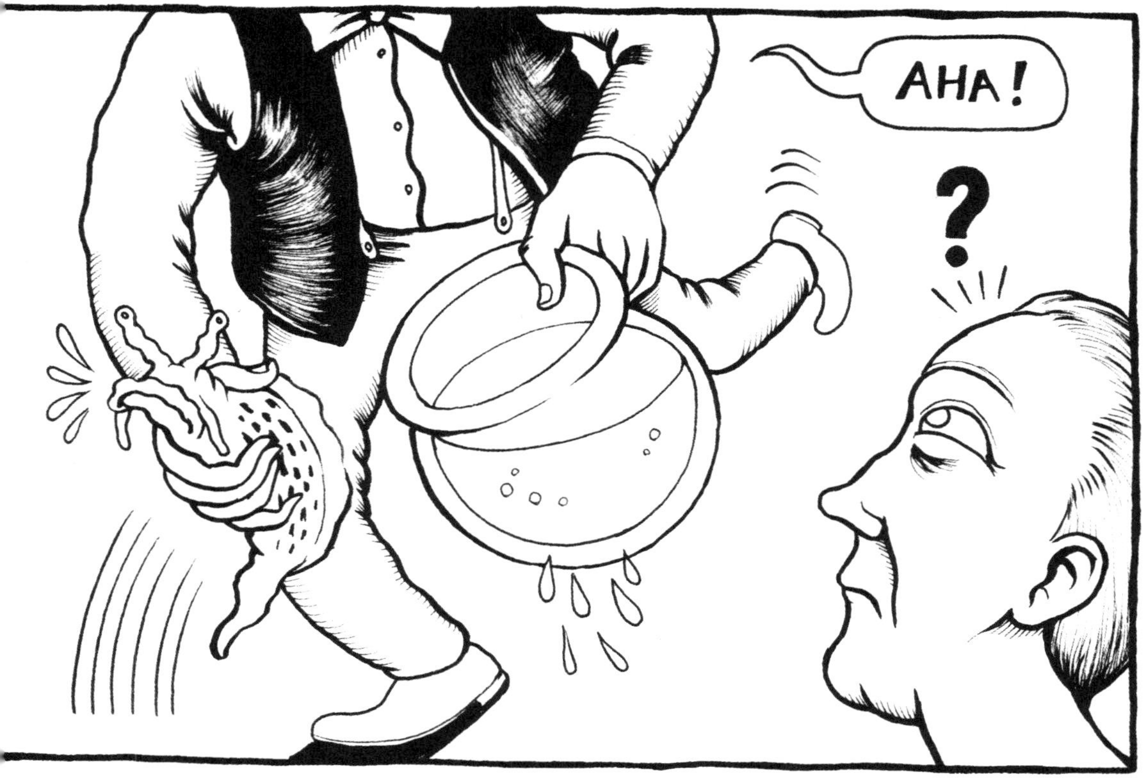
AHA!
?

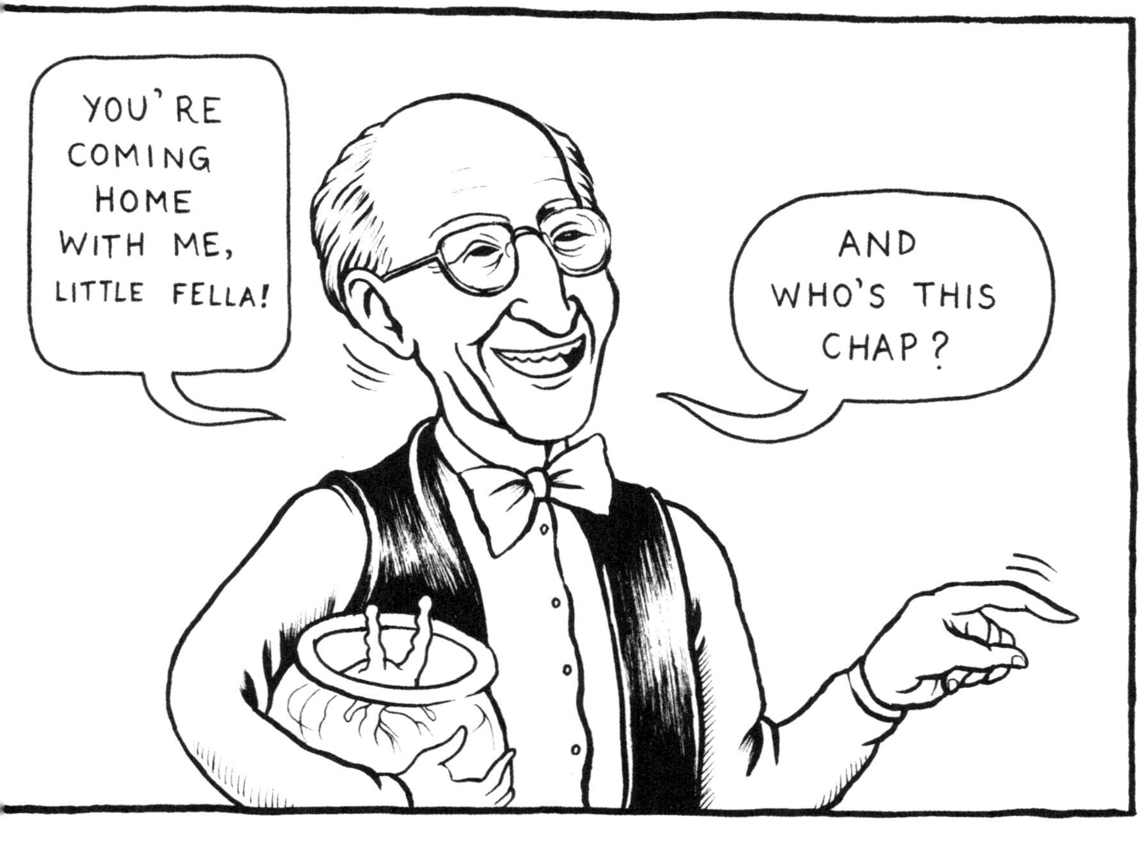
YOU'RE COMING HOME WITH ME, LITTLE FELLA!
AND WHO'S THIS CHAP?

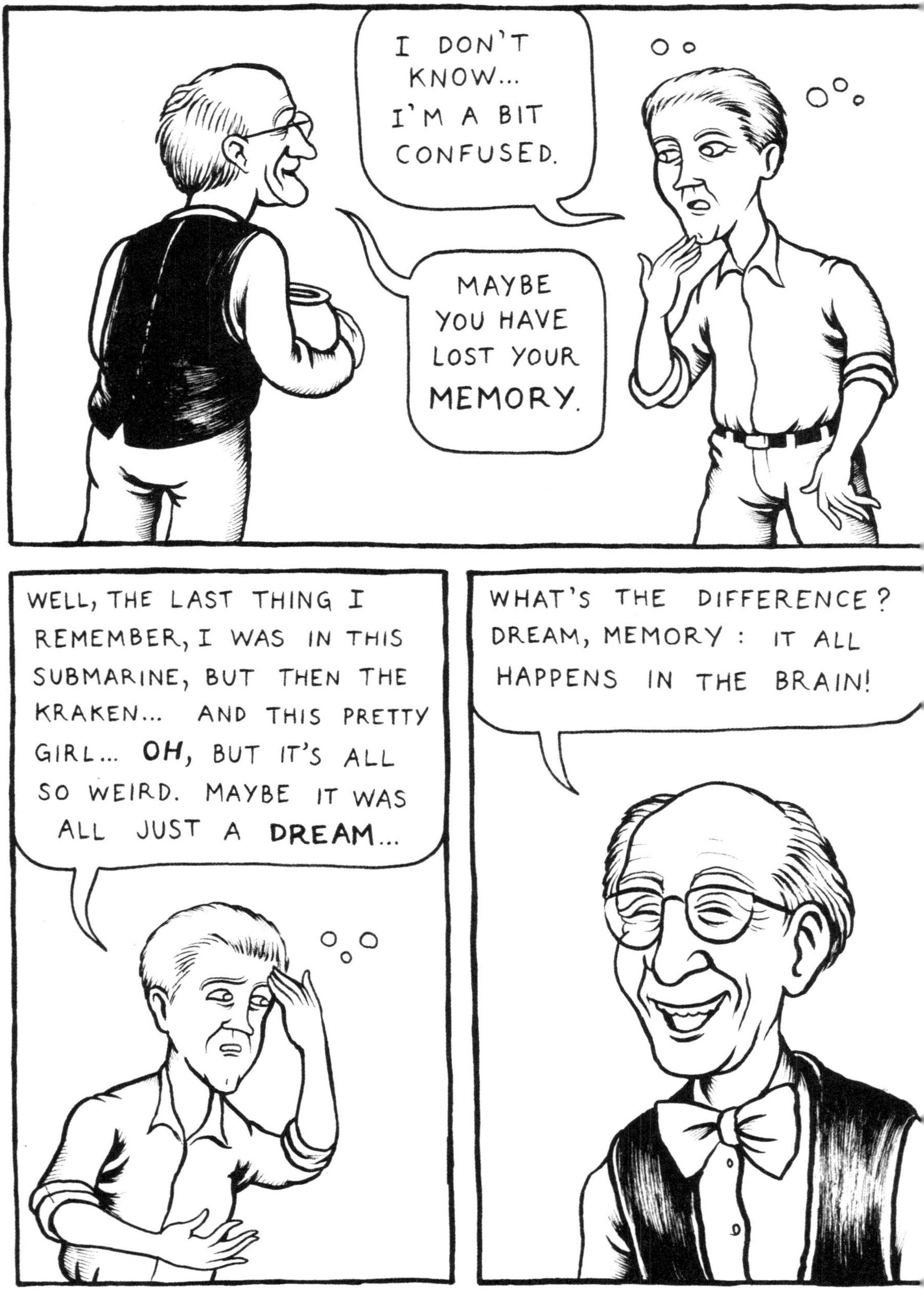
I DON'T KNOW... I'M A BIT CONFUSED.
MAYBE YOU HAVE LOST YOUR MEMORY.
WELL, THE LAST THING I REMEMBER, I WAS IN THIS SUBMARINE, BUT THEN THE KRAKEN... AND THIS PRETTY GIRL... OH, BUT IT'S ALL SO WEIRD. MAYBE IT WAS ALL JUST A DREAM...
WHAT'S THE DIFFERENCE? DREAM, MEMORY: IT ALL HAPPENS IN THE BRAIN!

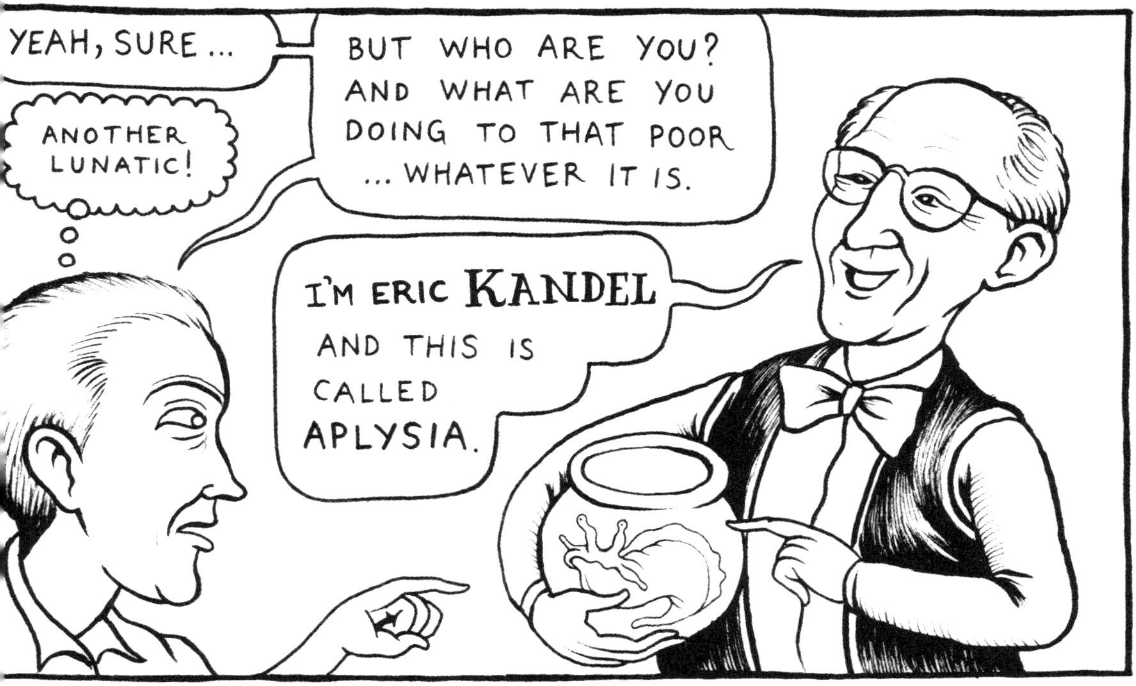
YEAH, SURE...
ANOTHER LUNATIC!
BUT WHO ARE YOU? AND WHAT ARE YOU DOING TO THAT POOR ... WHATEVER IT IS.
I'M ERIC KANDEL AND THIS IS CALLED APLYSIA.

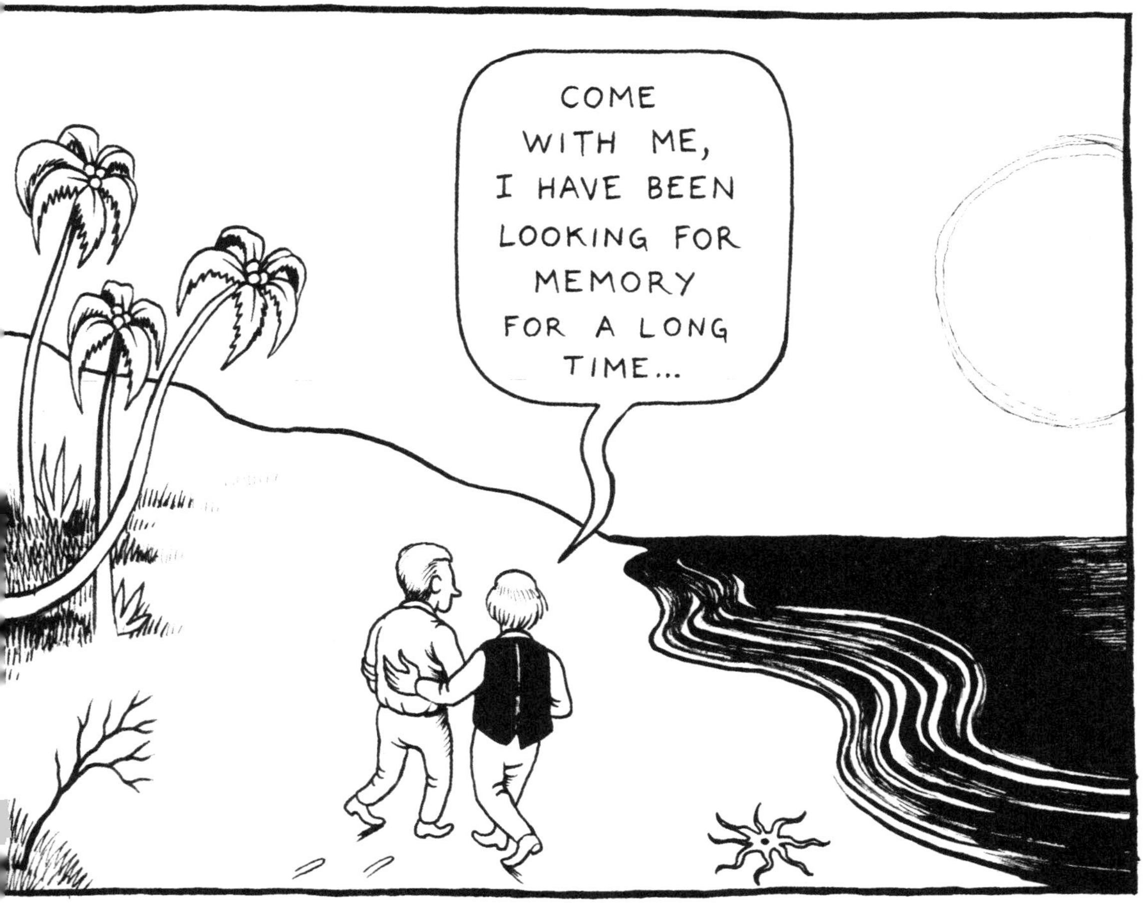
COME WITH ME, I HAVE BEEN LOOKING FOR MEMORY FOR A LONG TIME...

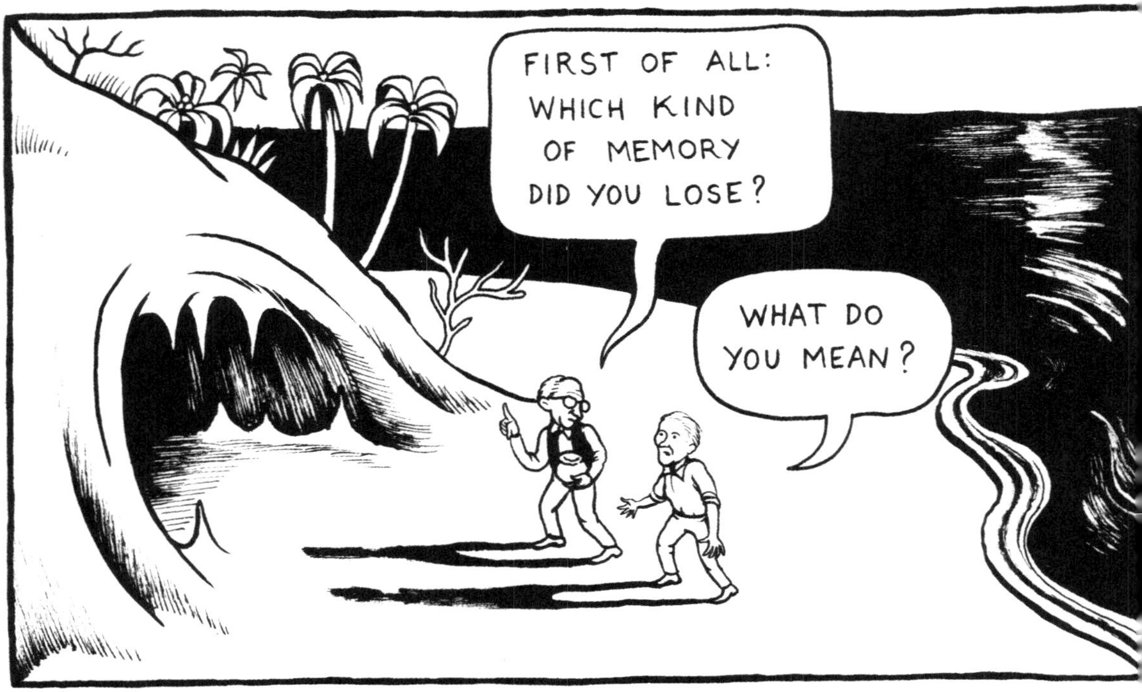
FIRST OF ALL: WHICH KIND OF MEMORY DID YOU LOSE?
WHAT DO YOU MEAN?

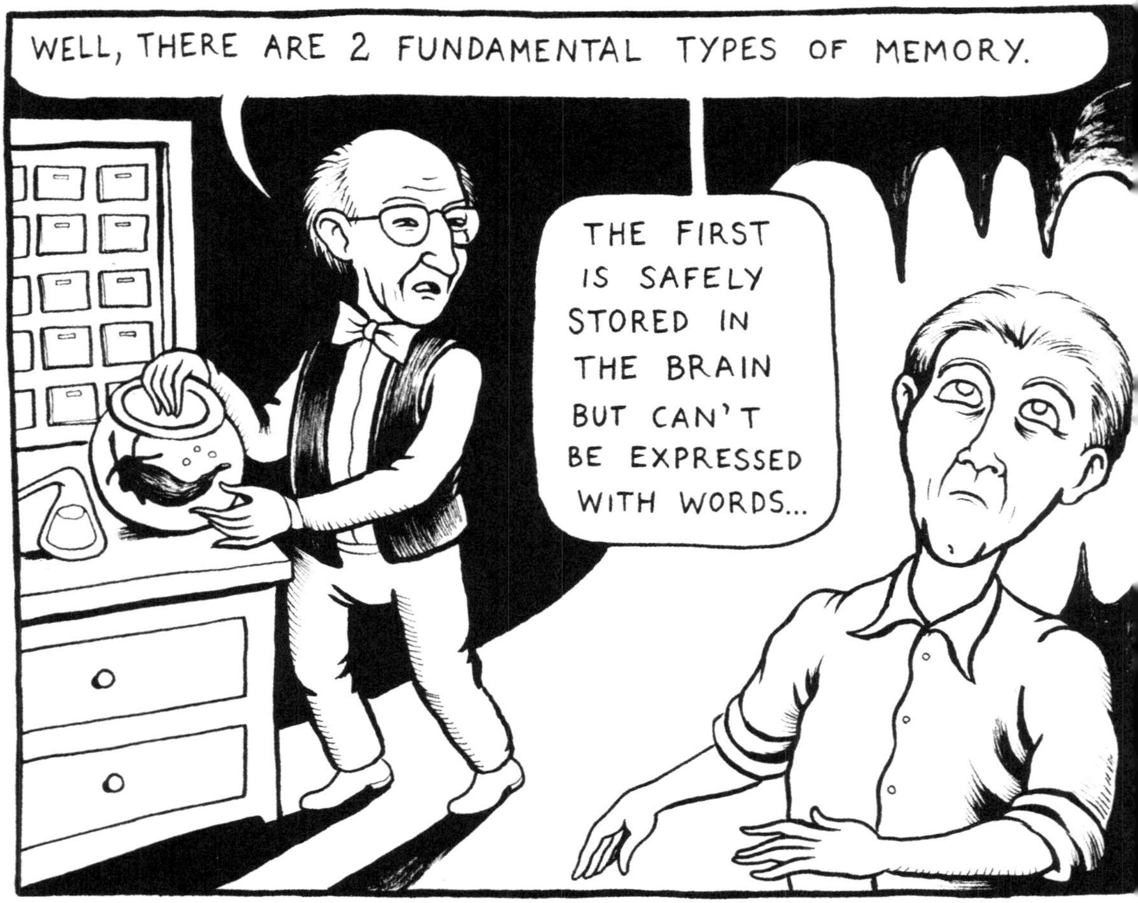
WELL, THERE ARE 2 FUNDAMENTAL TYPES OF MEMORY.
THE FIRST IS SAFELY STORED IN THE BRAIN BUT CAN'T BE EXPRESSED WITH WORDS...

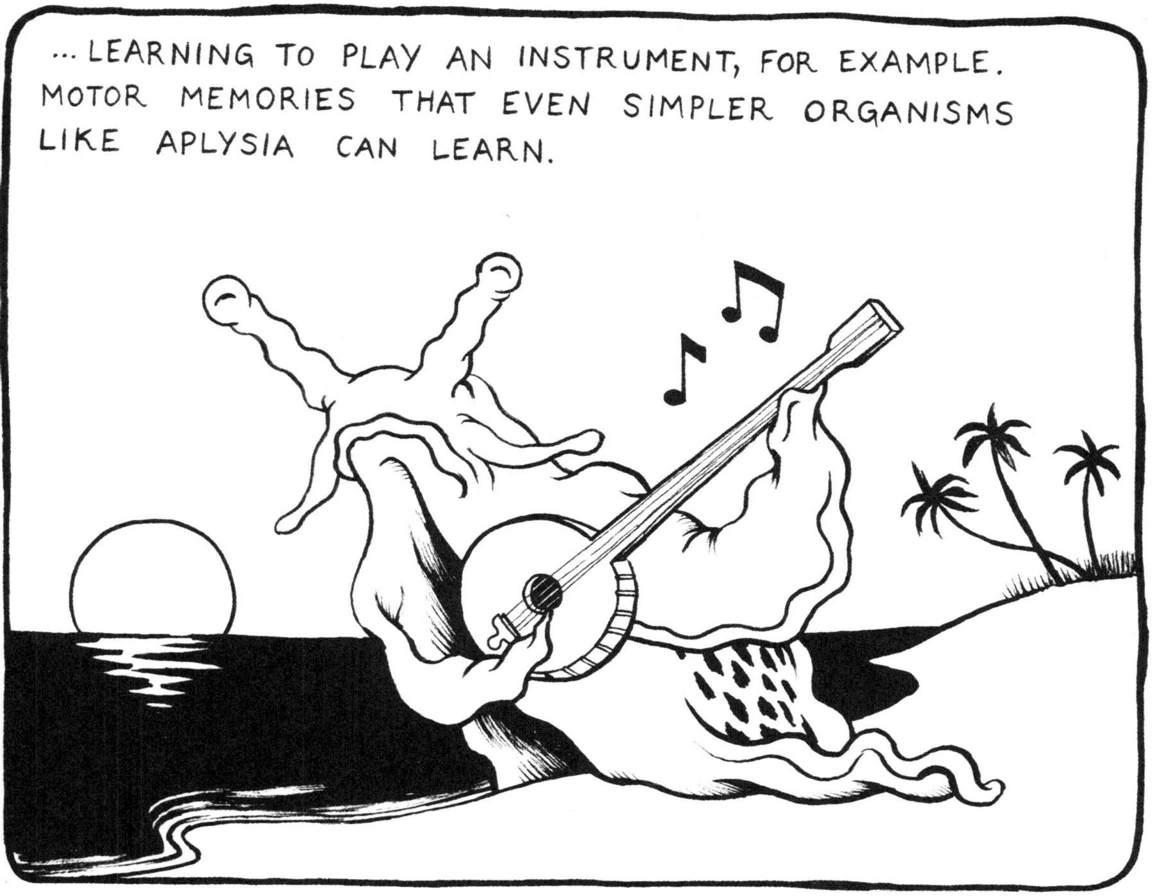
...LEARNING TO PLAY AN INSTRUMENT, FOR EXAMPLE. MOTOR MEMORIES THAT EVEN SIMPLER ORGANISMS LIKE APLYSIA CAN LEARN.

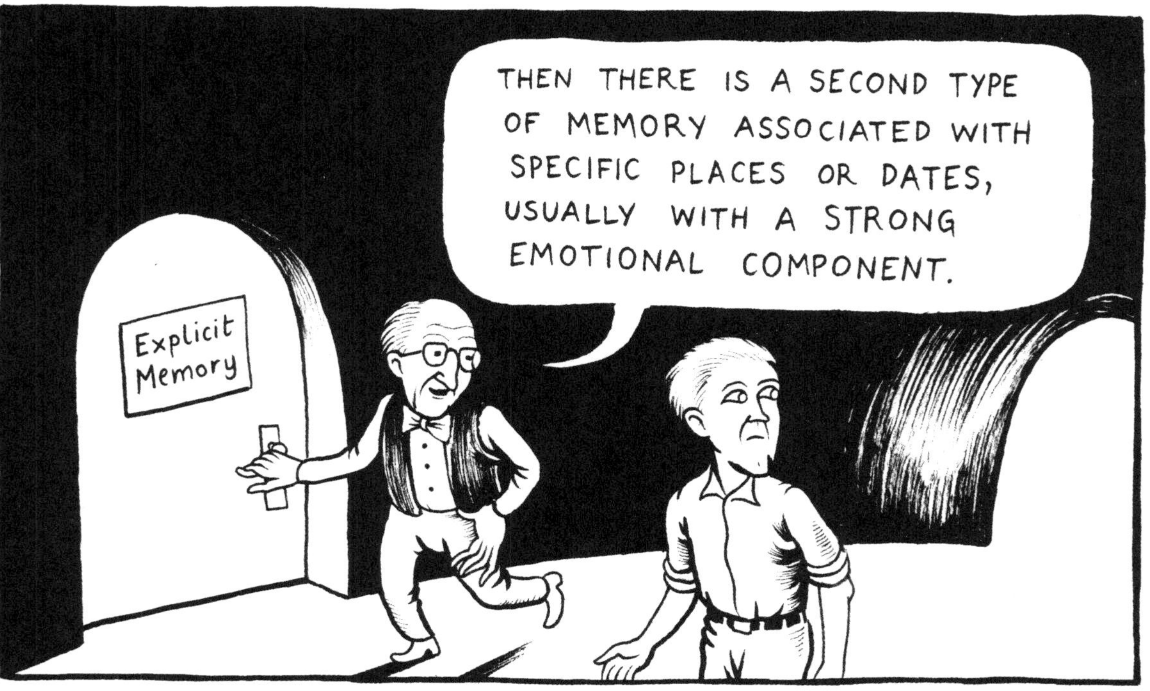
THEN THERE IS A SECOND TYPE OF MEMORY ASSOCIATED WITH SPECIFIC PLACES OR DATES, USUALLY WITH A STRONG EMOTIONAL COMPONENT.
Explicit Memory

THIS SECOND TYPE OF MEMORY IS RECORDED BY THE HIPPOCAMPUS, THE ARCHIVIST OF THE BRAIN. IT HAS TO DEAL WITH ALL THE DATES, THE MAPS AND THE IMPORTANT FACTS OF YOUR LIFE...
IT'S REALLY HARD WORK.
MMH, MAPS?

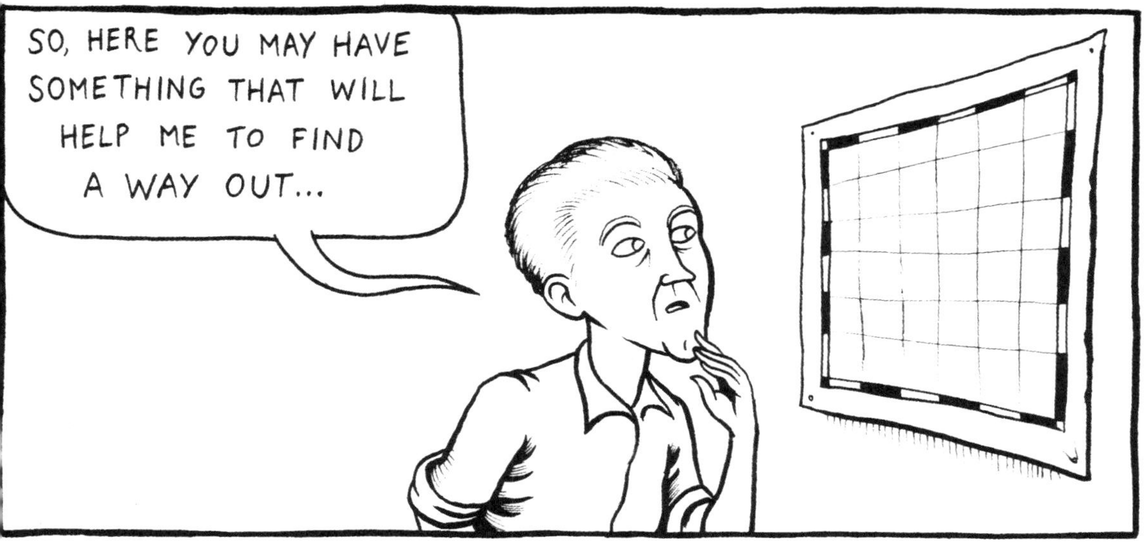
SO, HERE YOU MAY HAVE SOMETHING THAT WILL HELP ME TO FIND A WAY OUT...

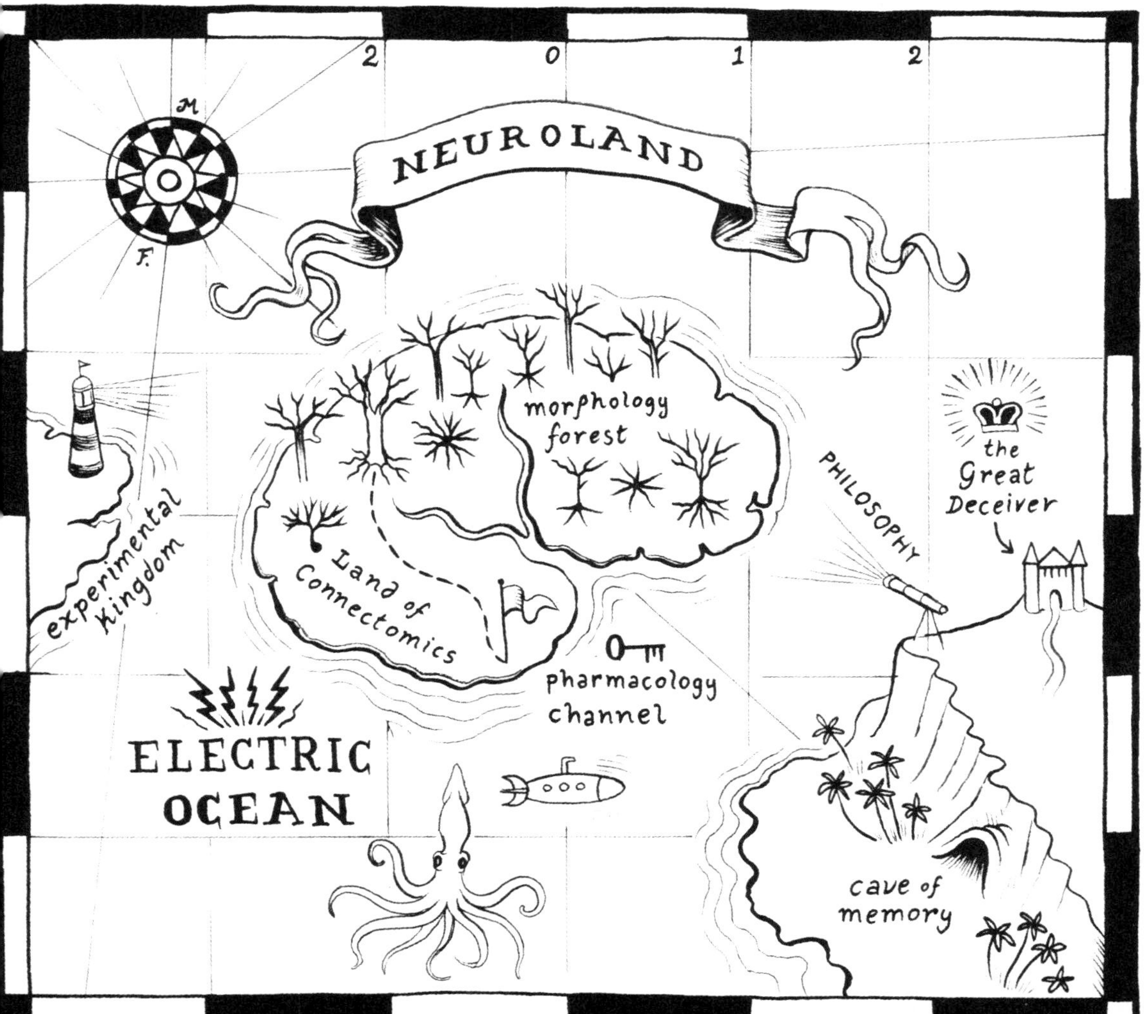
2 0 1 2
M
F.
NEUROLAND
morphology forest
Land of Connectomics
experimental kingdom
PHILOSOPHY
the Great Deceiver
pharmacology channel
ELECTRIC OCEAN
cave of memory

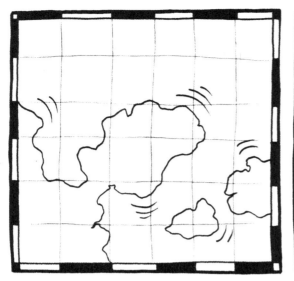

THIS STRANGE MAP IS CHANGING ALL THE TIME! HOW AM I SUPPOSED TO READ IT?

OF COURSE! THIS IS THE GREAT POWER OF THE BRAIN: IT'S **PLASTIC**! ONCE YOU LEARN SOMETHING IT IS NOT SET IN STONE, IT'S CONTINUOUSLY SHAPED BY **EXPERIENCE**.

BUT I HAVE TO FIND A WAY OUT OF HERE!

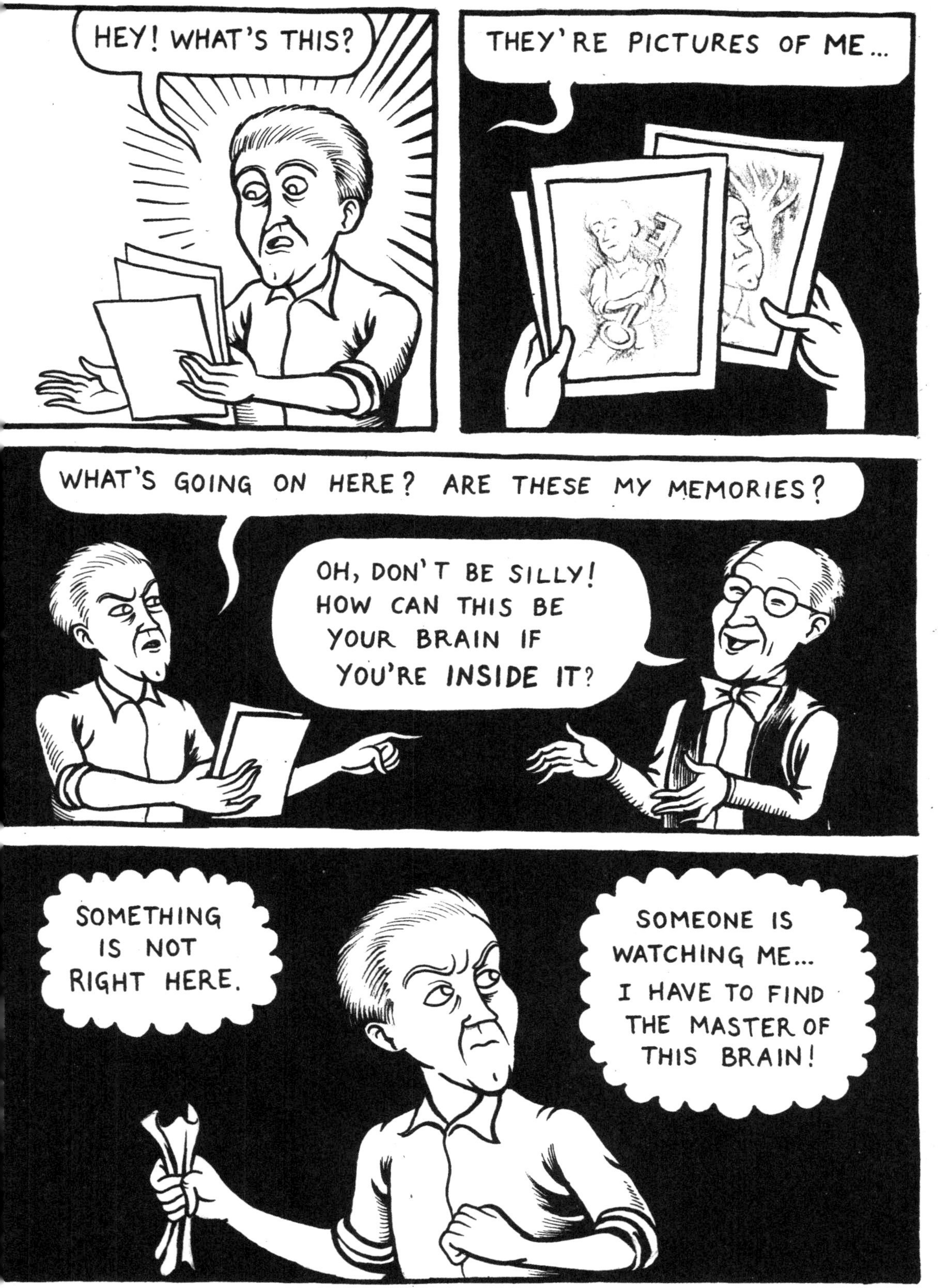
HEY! WHAT'S THIS?
THEY'RE PICTURES OF ME...
WHAT'S GOING ON HERE? ARE THESE MY MEMORIES?
OH, DON'T BE SILLY! HOW CAN THIS BE YOUR BRAIN IF YOU'RE INSIDE IT?
SOMETHING IS NOT RIGHT HERE.
SOMEONE IS WATCHING ME... I HAVE TO FIND THE MASTER OF THIS BRAIN!

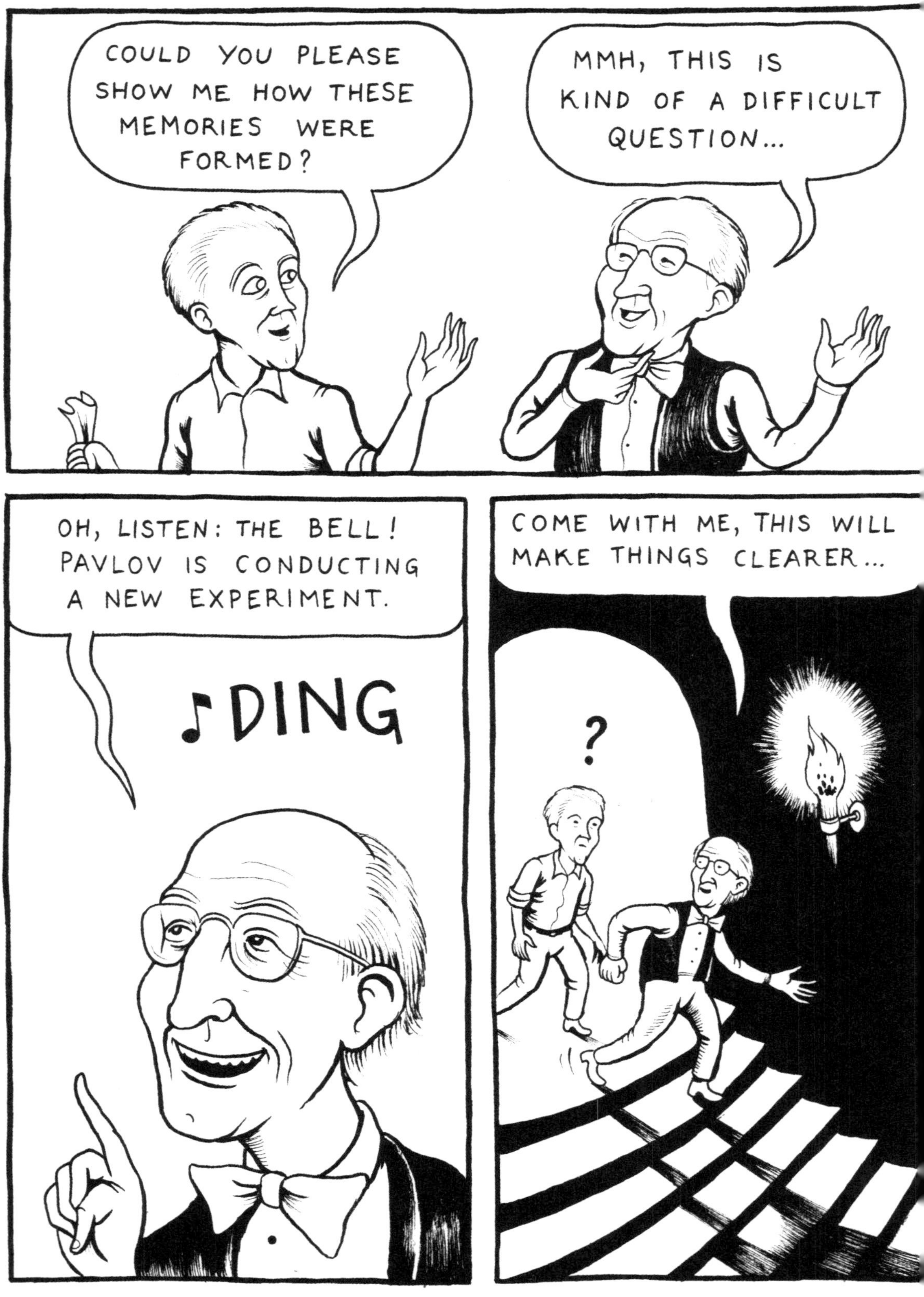
COULD YOU PLEASE SHOW ME HOW THESE MEMORIES WERE FORMED?
MMH, THIS IS KIND OF A DIFFICULT QUESTION...
OH, LISTEN: THE BELL! PAVLOV IS CONDUCTING A NEW EXPERIMENT.
♪ DING
COME WITH ME, THIS WILL MAKE THINGS CLEARER...
?

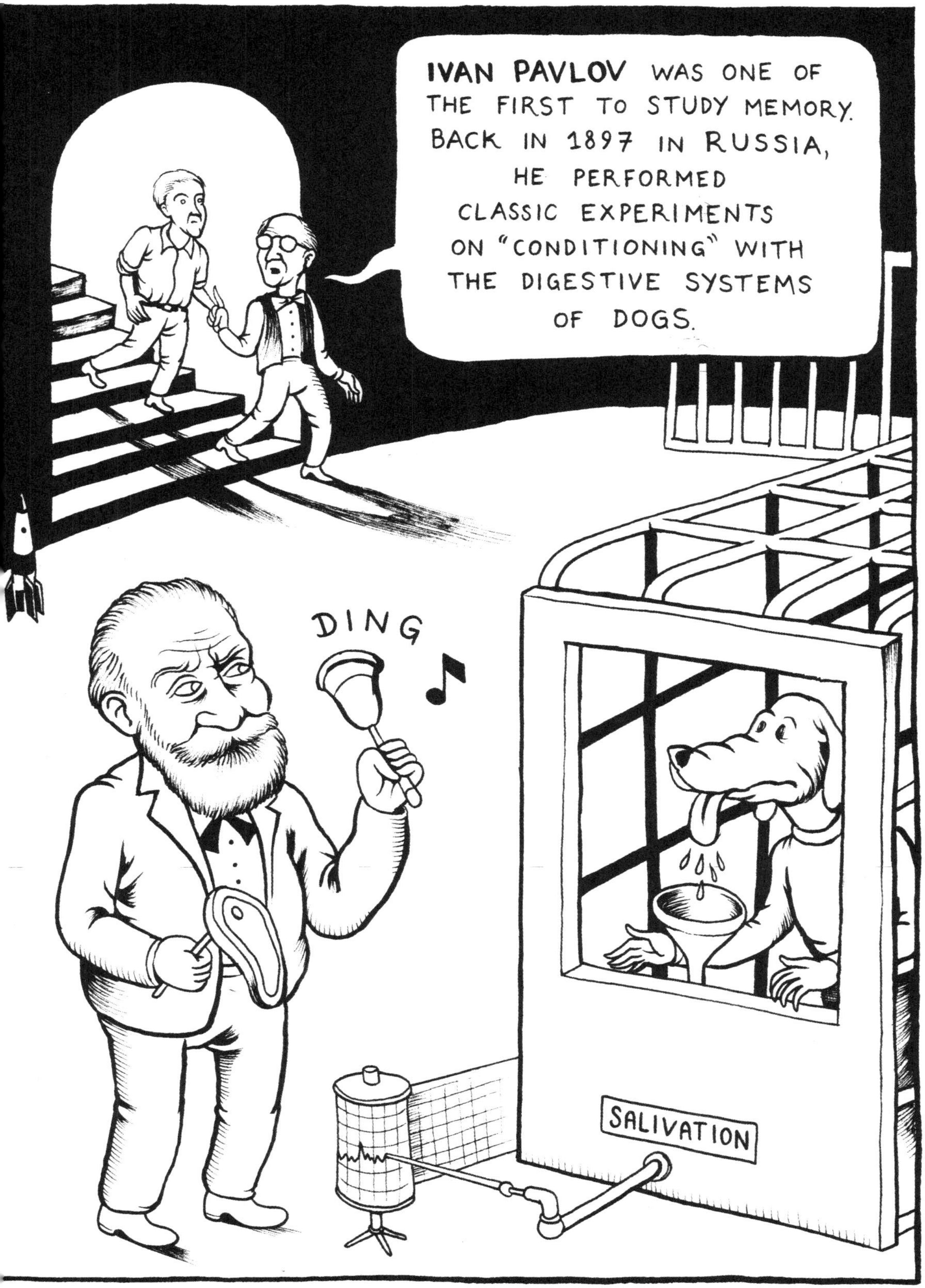
IVAN PAVLOV WAS ONE OF THE FIRST TO STUDY MEMORY. BACK IN 1897 IN RUSSIA, HE PERFORMED CLASSIC EXPERIMENTS ON "CONDITIONING" WITH THE DIGESTIVE SYSTEMS OF DOGS.
DING
SALIVATION

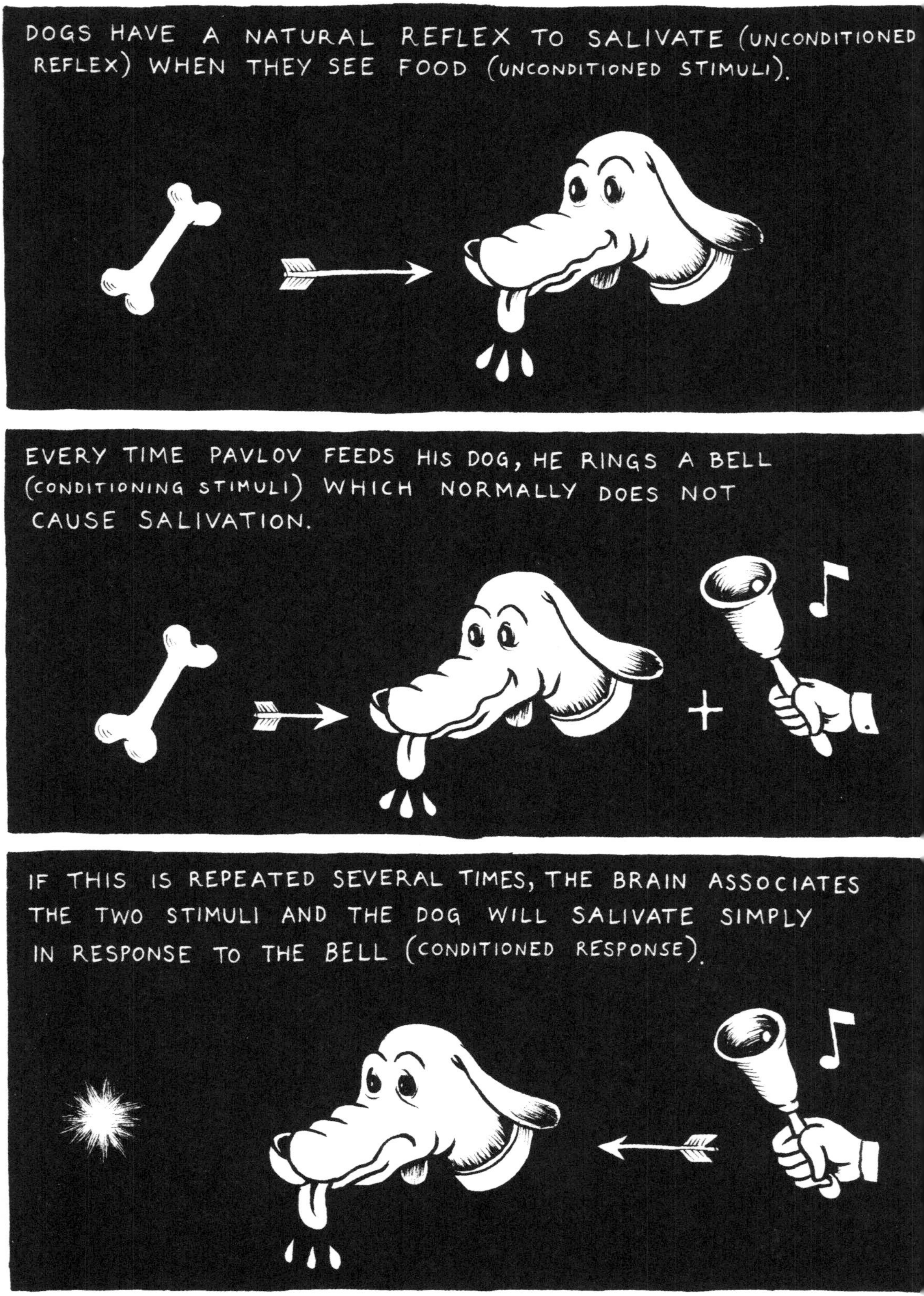
DOGS HAVE A NATURAL REFLEX TO SALIVATE (UNCONDITIONED REFLEX) WHEN THEY SEE FOOD (UNCONDITIONED STIMULI).
EVERY TIME PAVLOV FEEDS HIS DOG, HE RINGS A BELL (CONDITIONING STIMULI) WHICH NORMALLY DOES NOT CAUSE SALIVATION.
IF THIS IS REPEATED SEVERAL TIMES, THE BRAIN ASSOCIATES THE TWO STIMULI AND THE DOG WILL SALIVATE SIMPLY IN RESPONSE TO THE BELL (CONDITIONED RESPONSE).

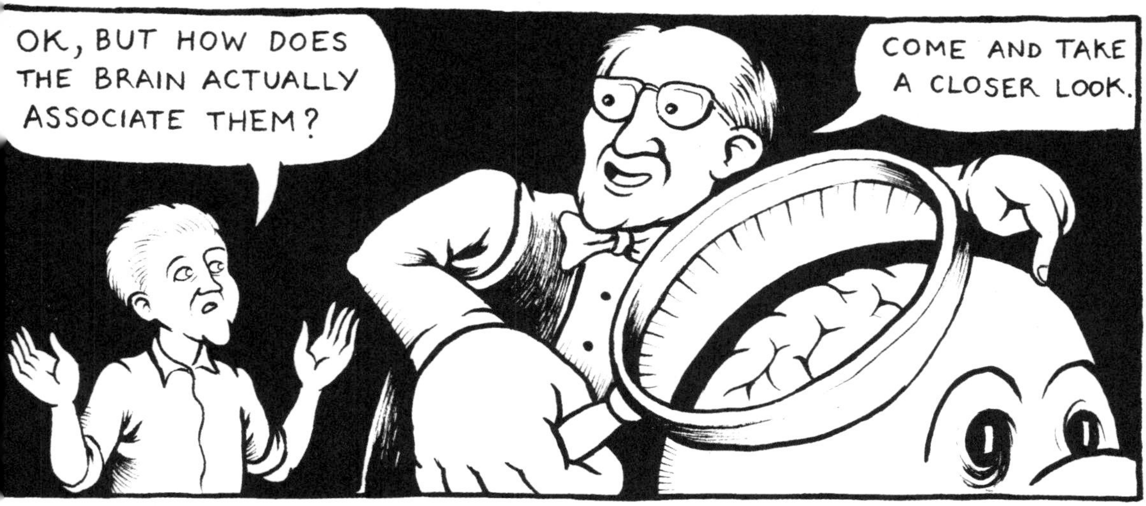
OK, BUT HOW DOES THE BRAIN ACTUALLY ASSOCIATE THEM?
COME AND TAKE A CLOSER LOOK.

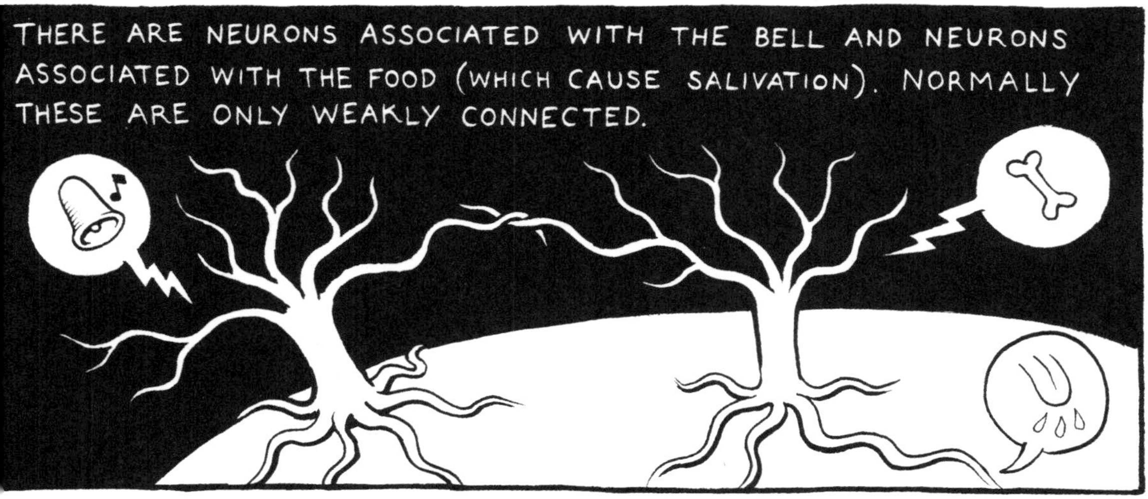
THERE ARE NEURONS ASSOCIATED WITH THE BELL AND NEURONS ASSOCIATED WITH THE FOOD (WHICH CAUSE SALIVATION). NORMALLY THESE ARE ONLY WEAKLY CONNECTED.

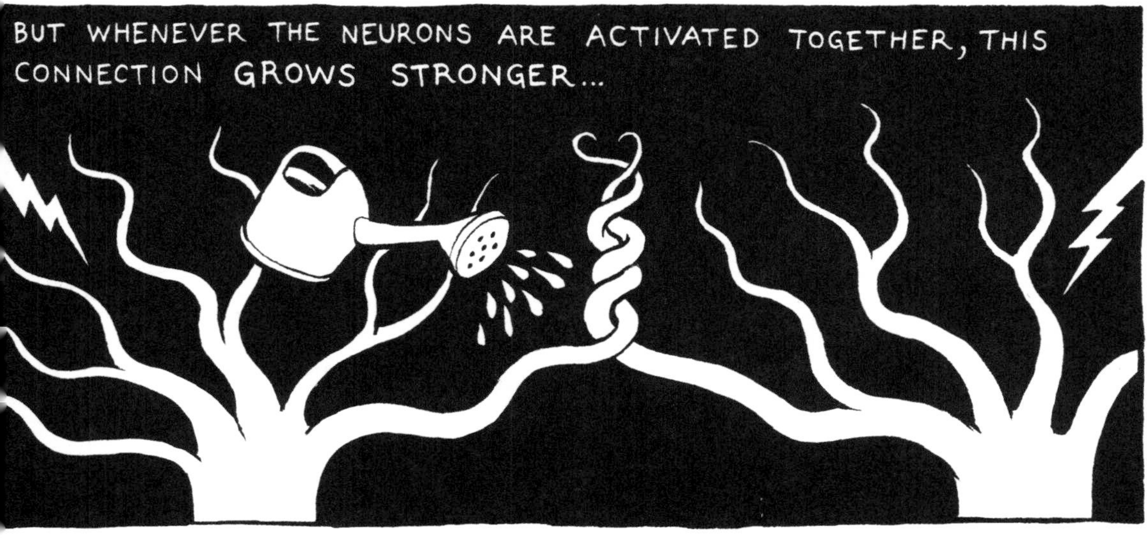
BUT WHENEVER THE NEURONS ARE ACTIVATED TOGETHER, THIS CONNECTION GROWS STRONGER...

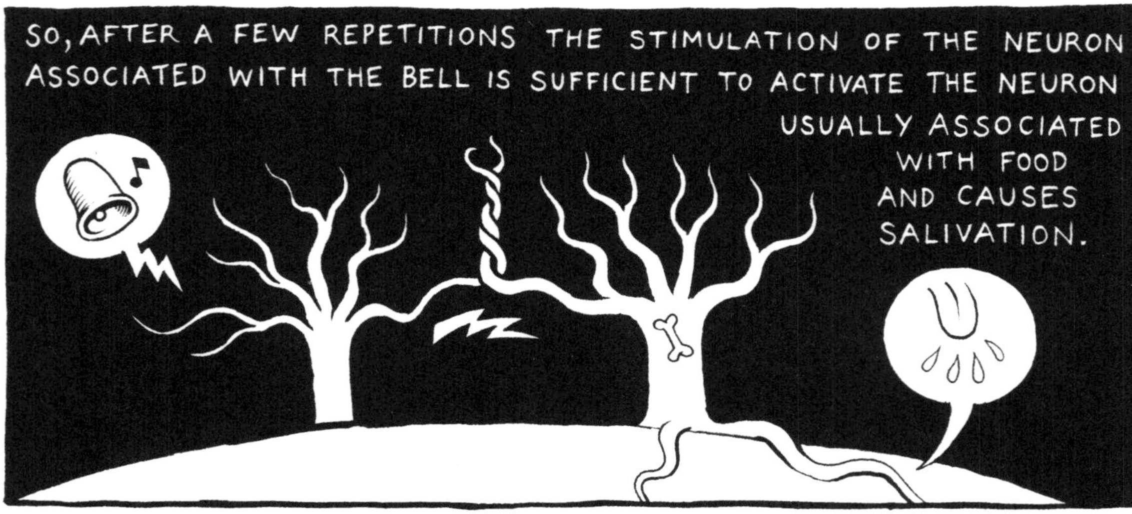
SO, AFTER A FEW REPETITIONS THE STIMULATION OF THE NEURON ASSOCIATED WITH THE BELL IS SUFFICIENT TO ACTIVATE THE NEURON USUALLY ASSOCIATED WITH FOOD AND CAUSES SALIVATION.

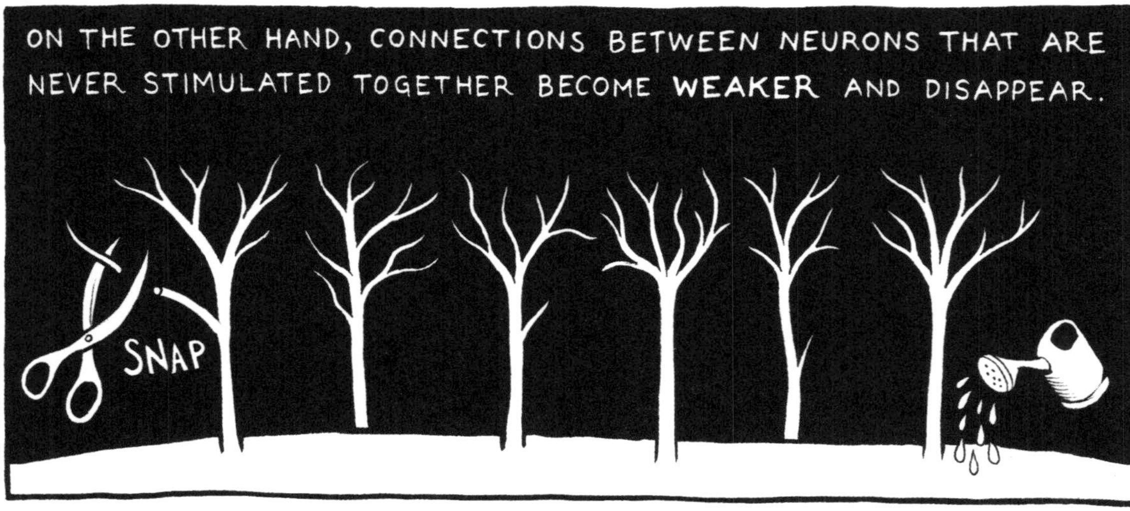
ON THE OTHER HAND, CONNECTIONS BETWEEN NEURONS THAT ARE NEVER STIMULATED TOGETHER BECOME WEAKER AND DISAPPEAR.
SNAP

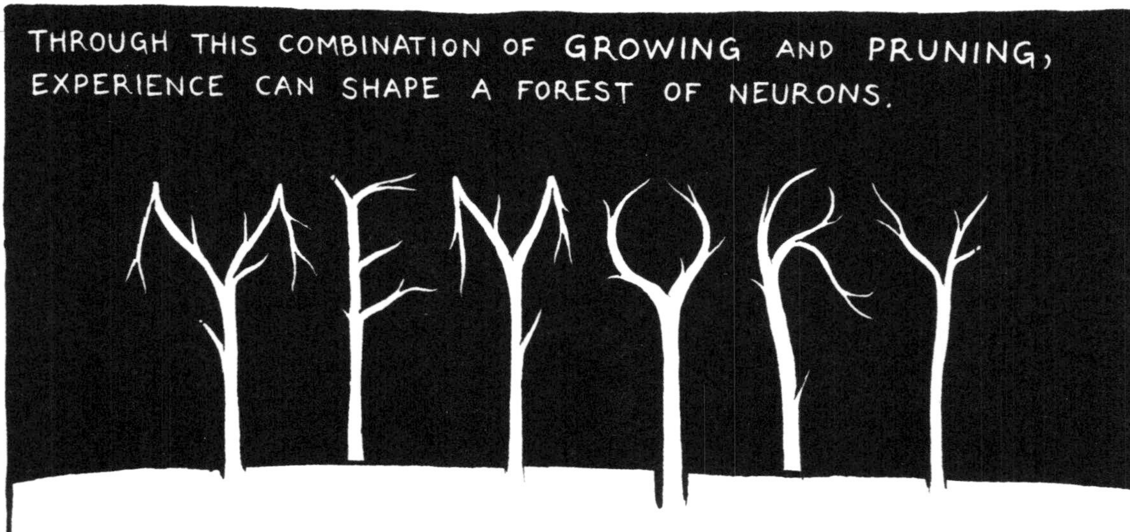
THROUGH THIS COMBINATION OF GROWING AND PRUNING, EXPERIENCE CAN SHAPE A FOREST OF NEURONS.
MEMORY

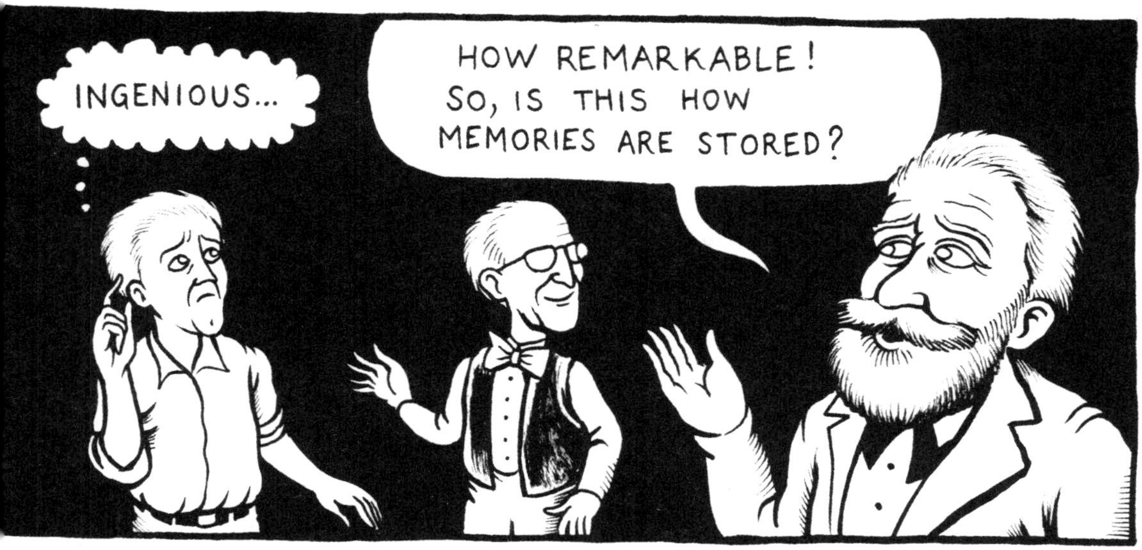
INGENIOUS...
HOW REMARKABLE! SO, IS THIS HOW MEMORIES ARE STORED?

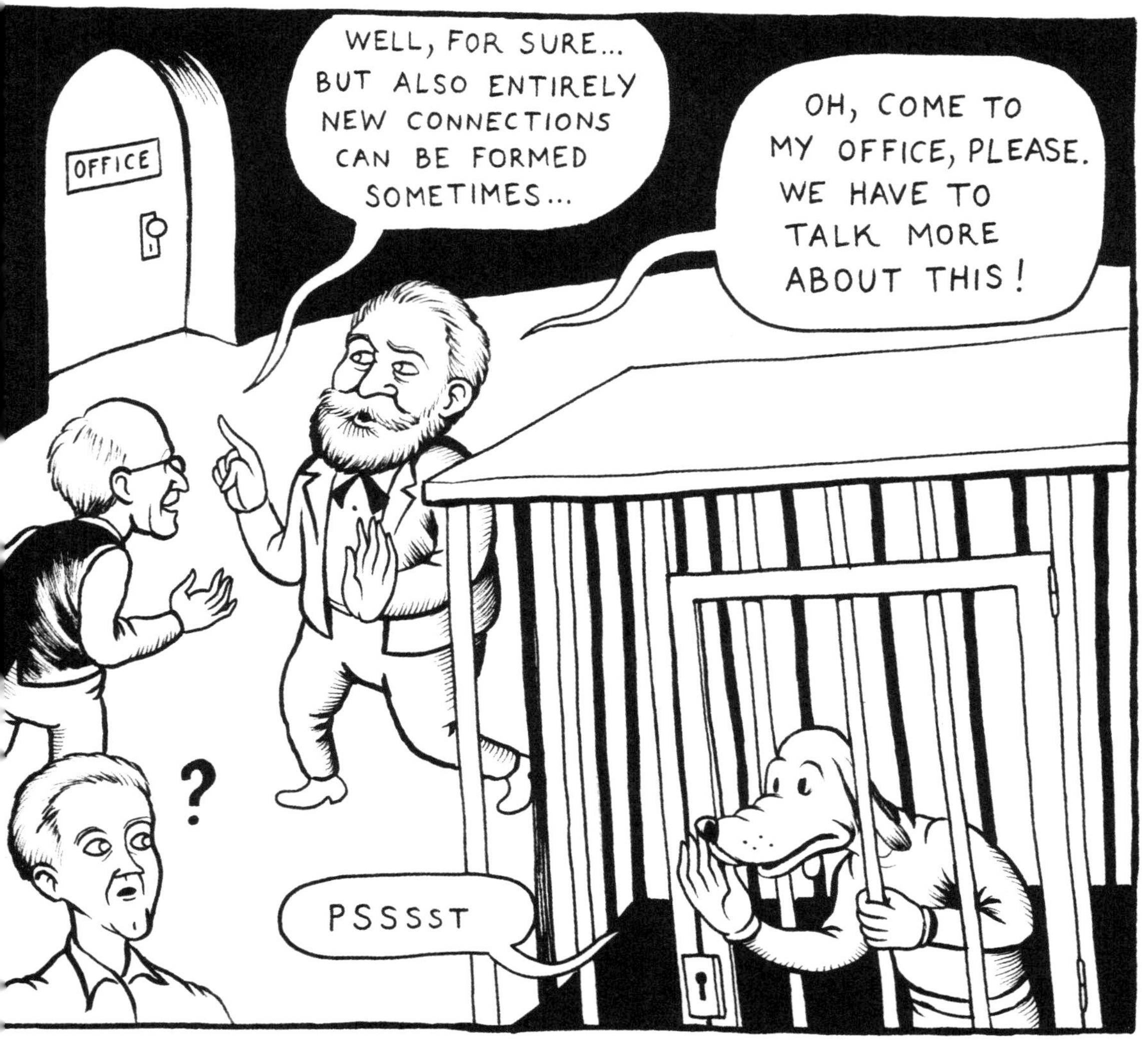
OFFICE
WELL, FOR SURE... BUT ALSO ENTIRELY NEW CONNECTIONS CAN BE FORMED SOMETIMES...
OH, COME TO MY OFFICE, PLEASE. WE HAVE TO TALK MORE ABOUT THIS!
?
PSSSST

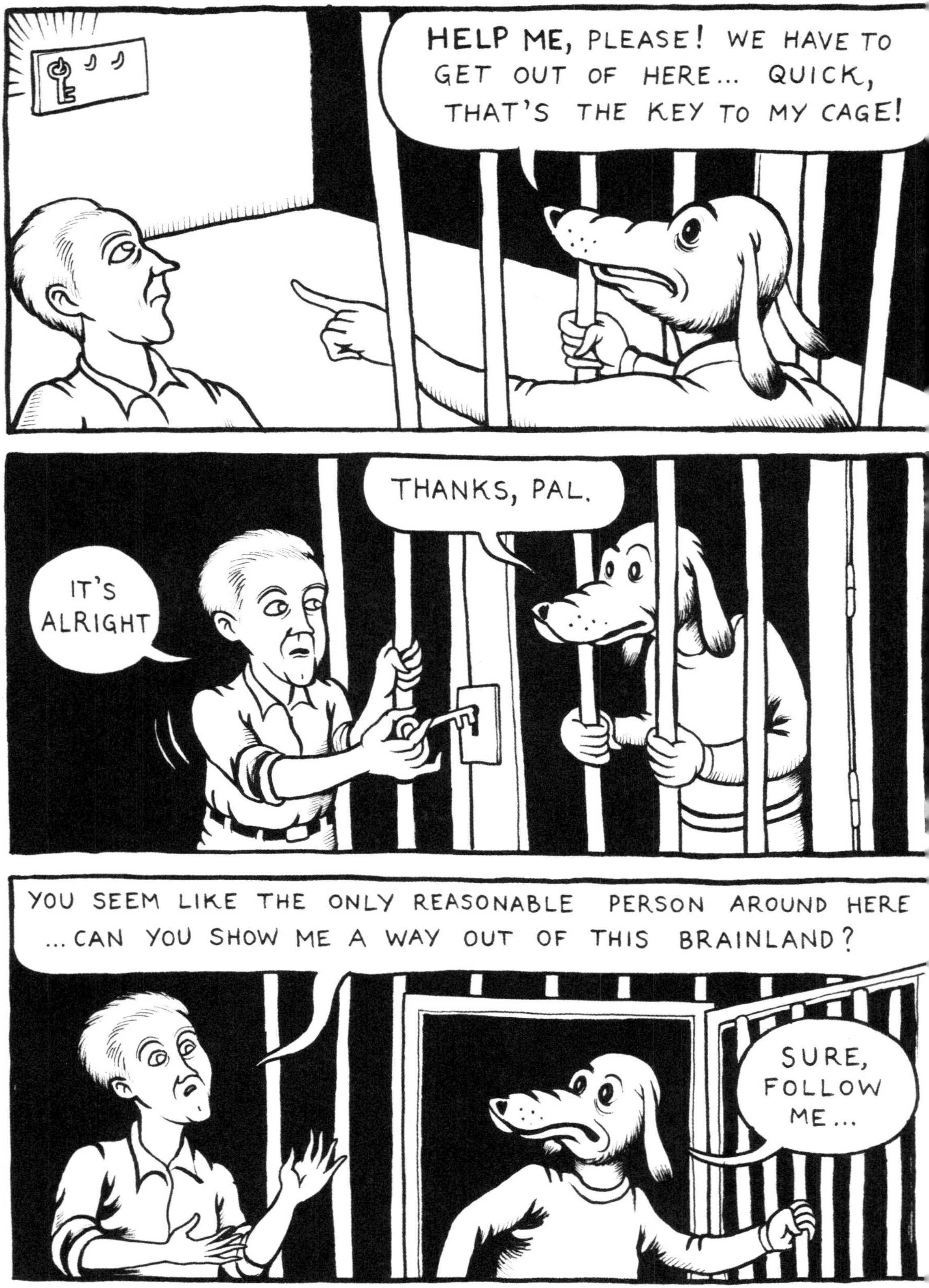
HELP ME, PLEASE! WE HAVE TO GET OUT OF HERE... QUICK, THAT'S THE KEY TO MY CAGE!
IT'S ALRIGHT
THANKS, PAL.
YOU SEEM LIKE THE ONLY REASONABLE PERSON AROUND HERE ...CAN YOU SHOW ME A WAY OUT OF THIS BRAINLAND?
SURE, FOLLOW ME...

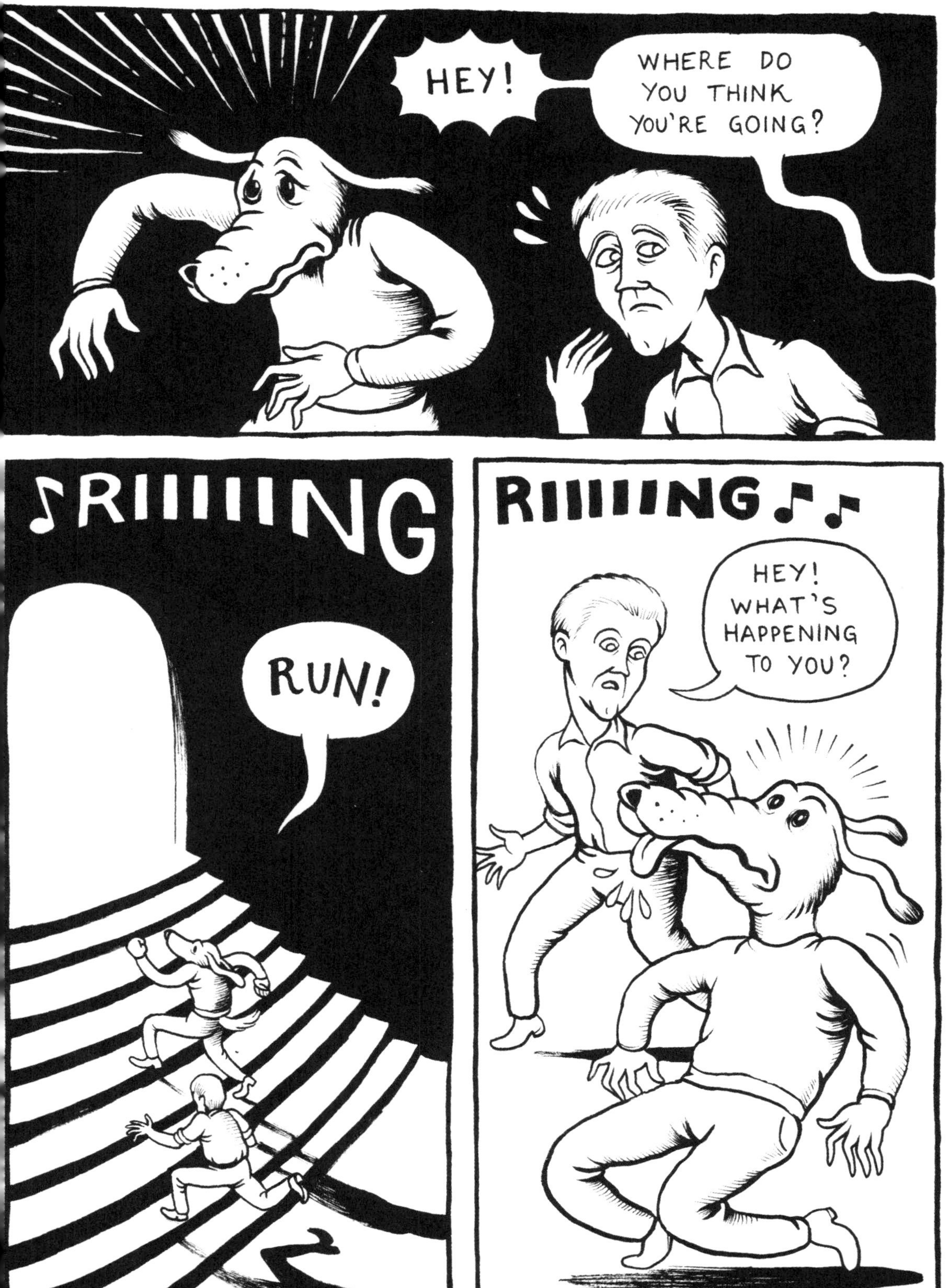
HEY!
WHERE DO YOU THINK YOU'RE GOING?
RIIIIING
RUN!
RIIIING
HEY! WHAT'S HAPPENING TO YOU?

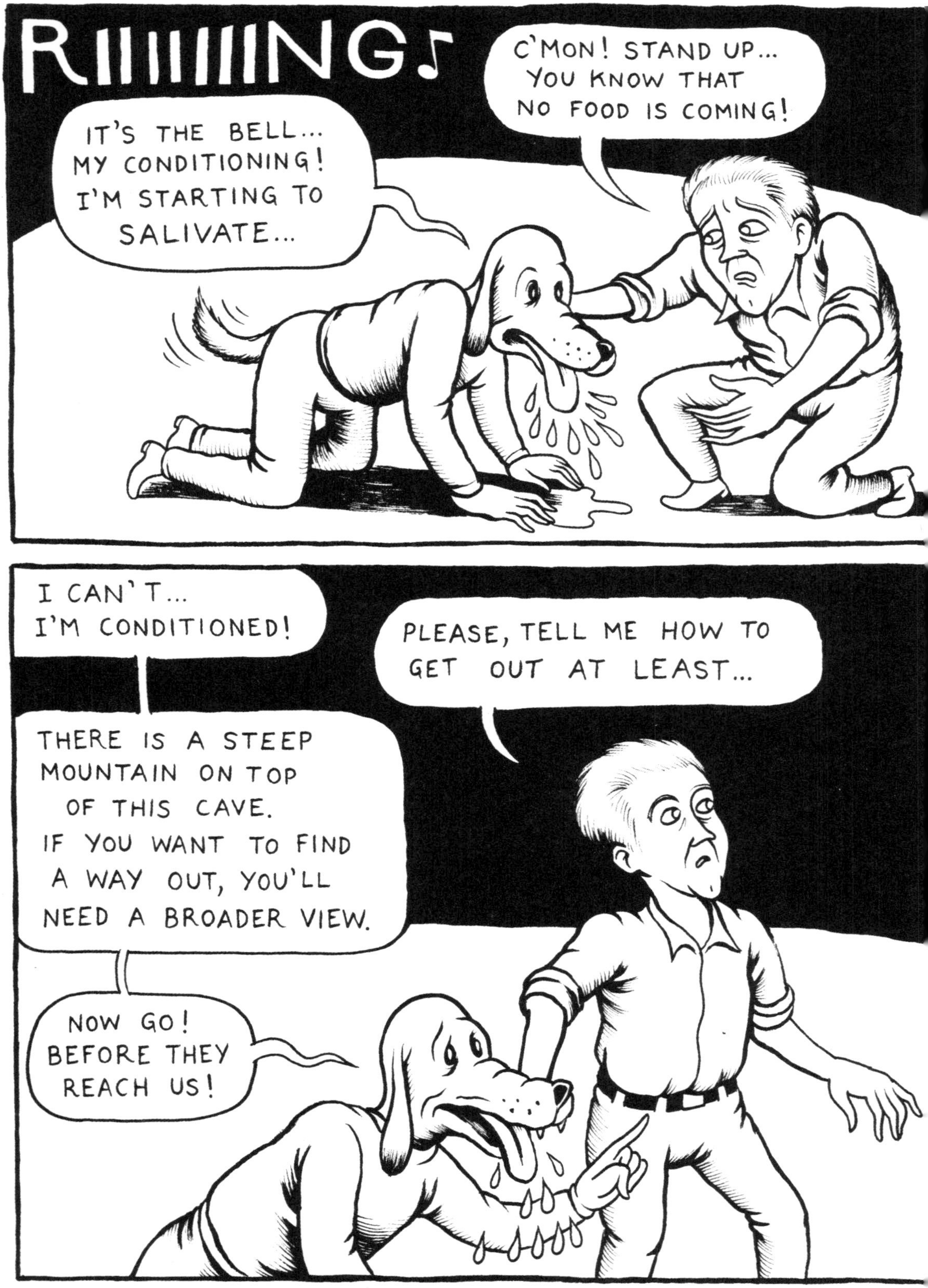
RIIIIIING♪
IT'S THE BELL... MY CONDITIONING! I'M STARTING TO SALIVATE...
C'MON! STAND UP... YOU KNOW THAT NO FOOD IS COMING!
I CAN'T... I'M CONDITIONED!
PLEASE, TELL ME HOW TO GET OUT AT LEAST...
THERE IS A STEEP MOUNTAIN ON TOP OF THIS CAVE. IF YOU WANT TO FIND A WAY OUT, YOU'LL NEED A BROADER VIEW.
NOW GO! BEFORE THEY REACH US!

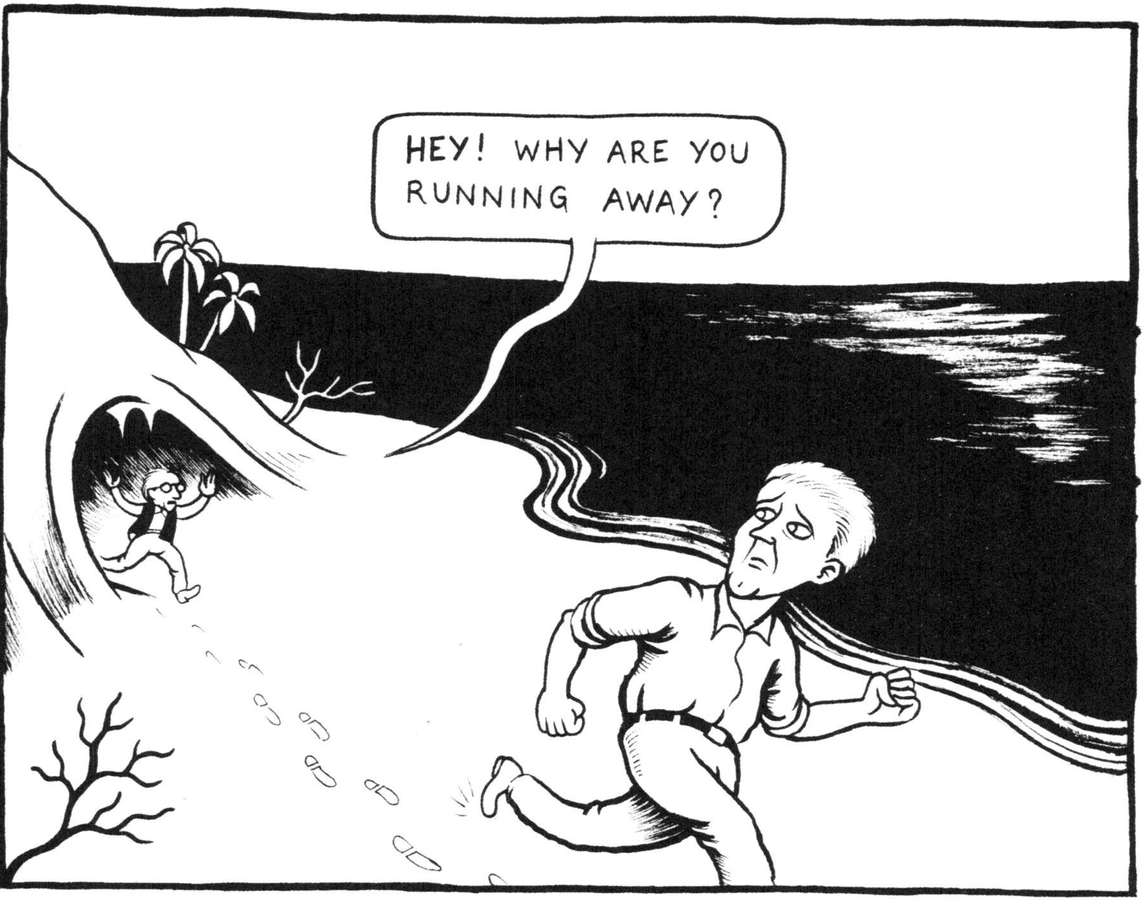
HEY! WHY ARE YOU RUNNING AWAY?

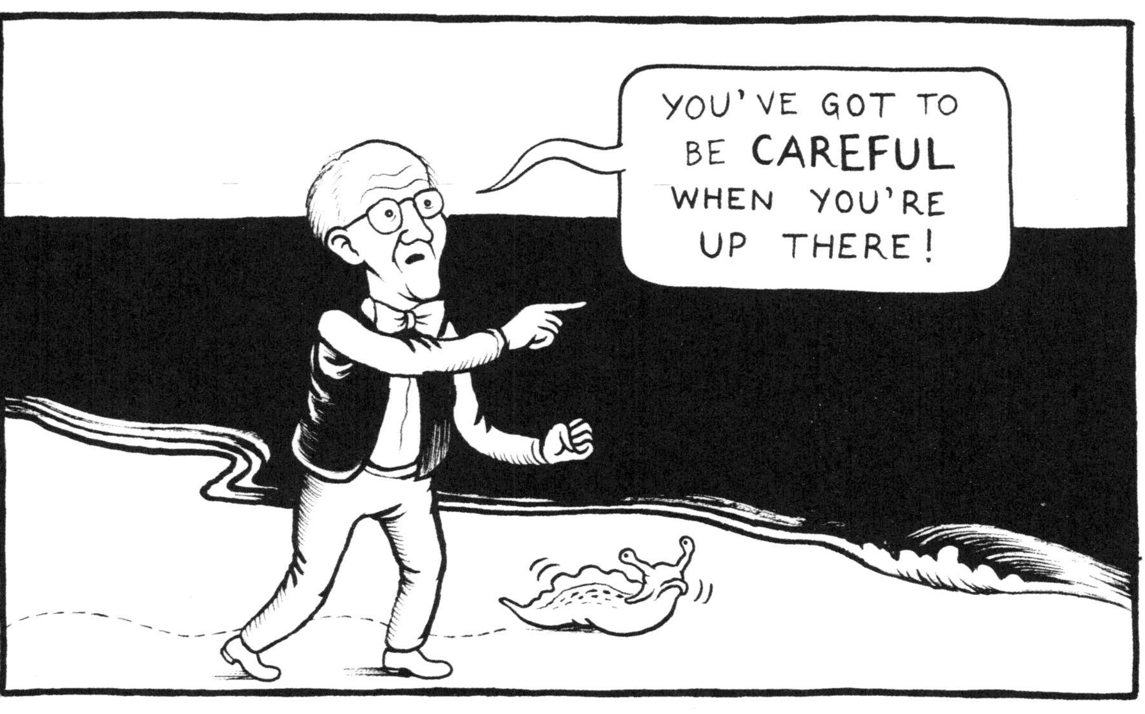
YOU'VE GOT TO BE CAREFUL WHEN YOU'RE UP THERE!

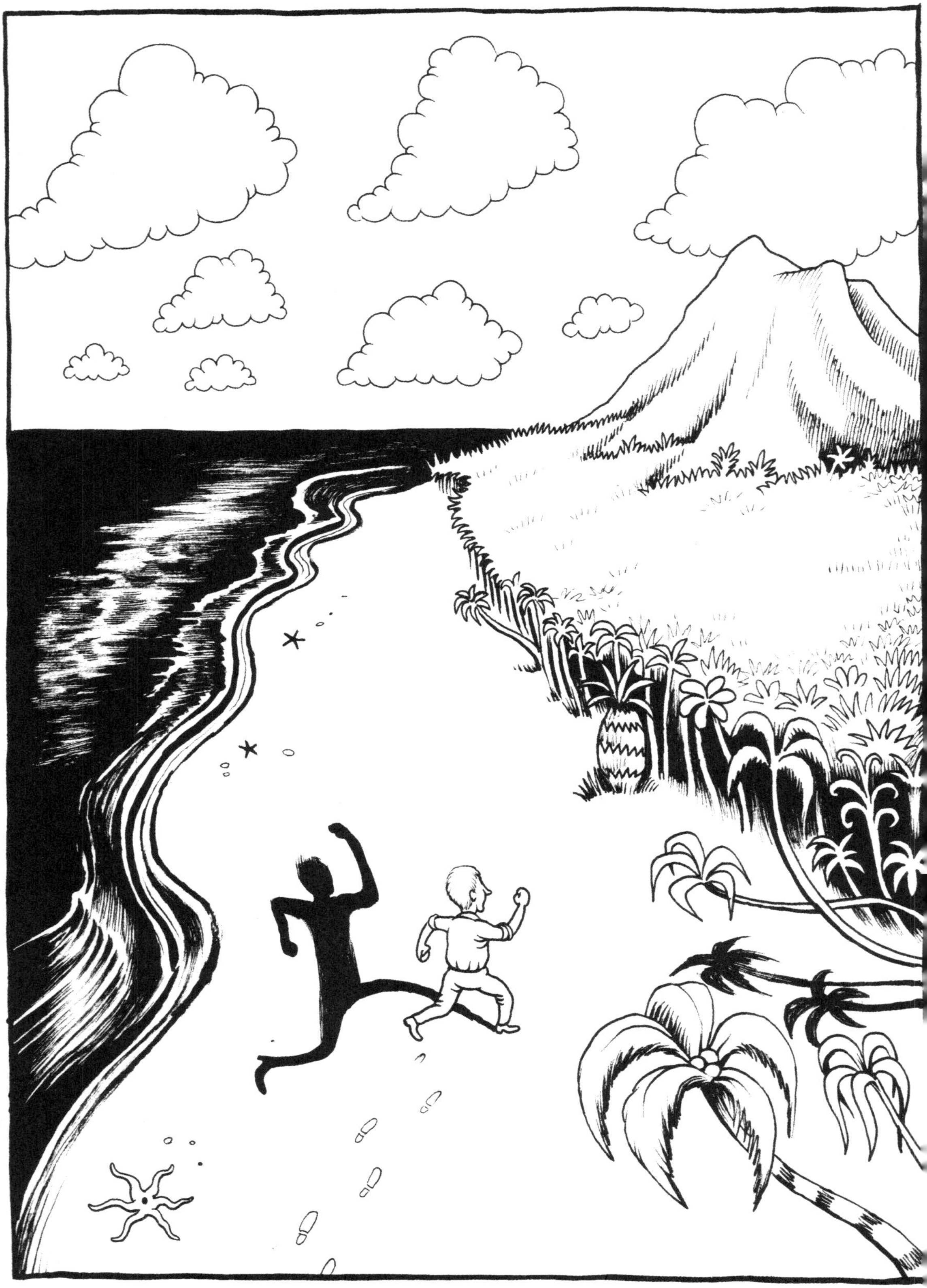

SYNCHRONICITY

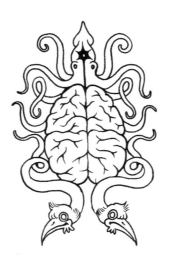

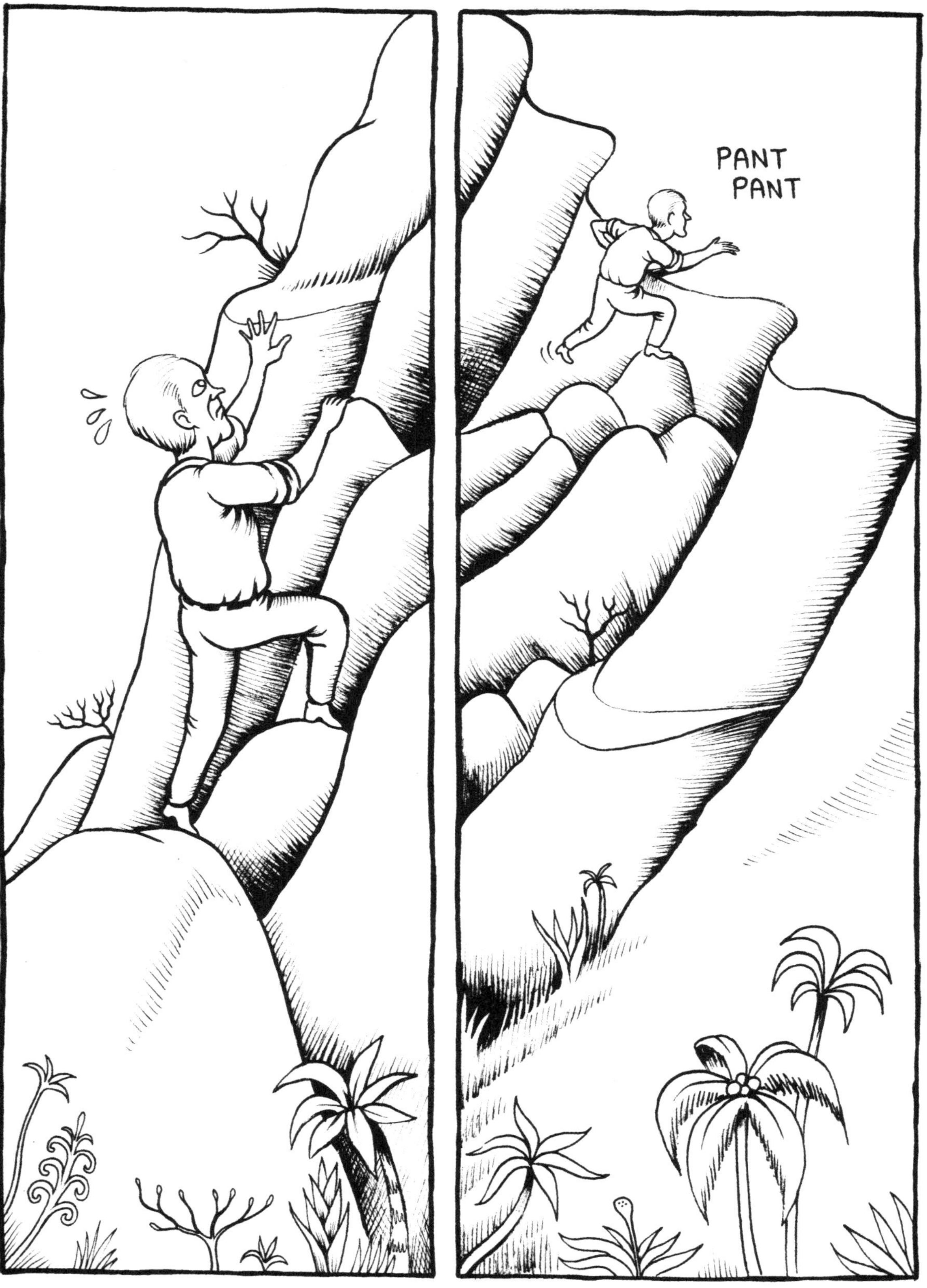
PANT
PANT

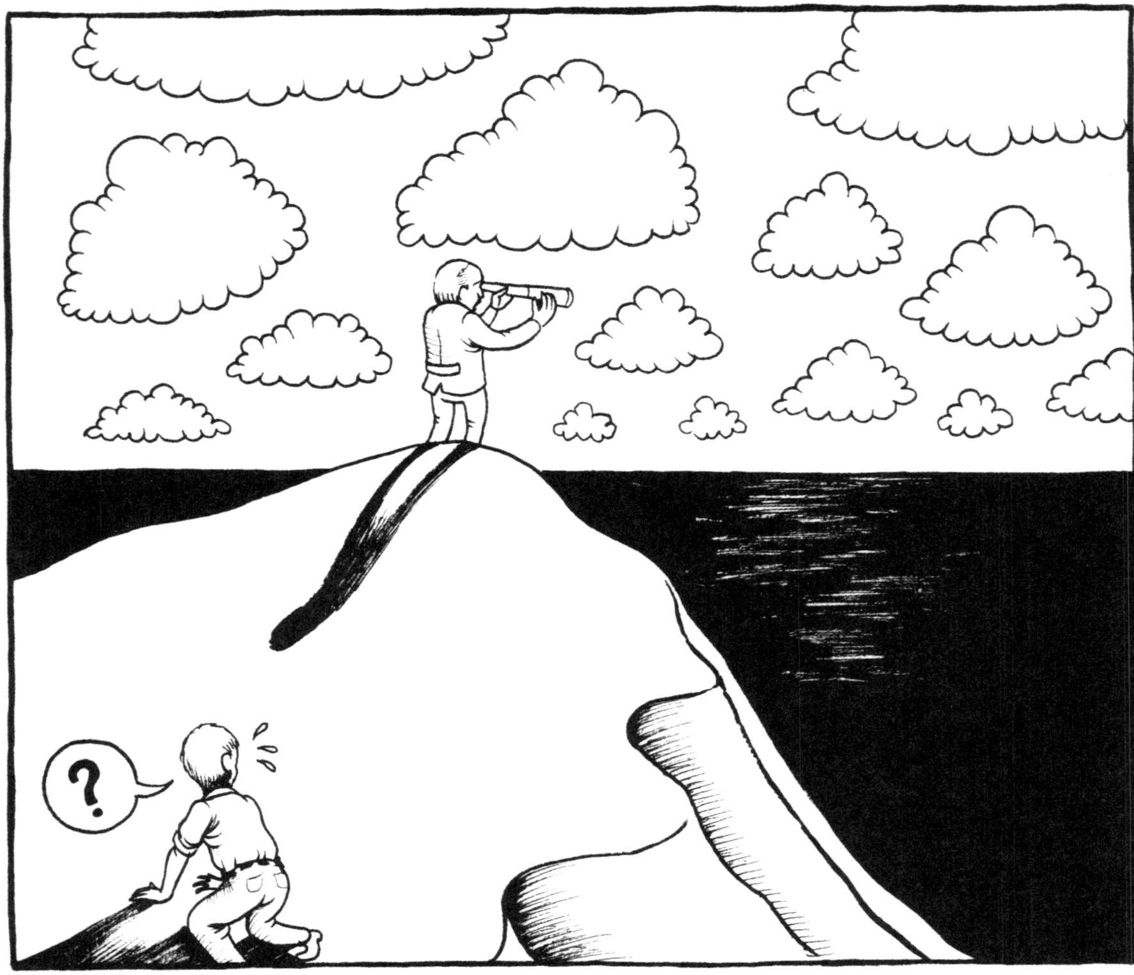
?

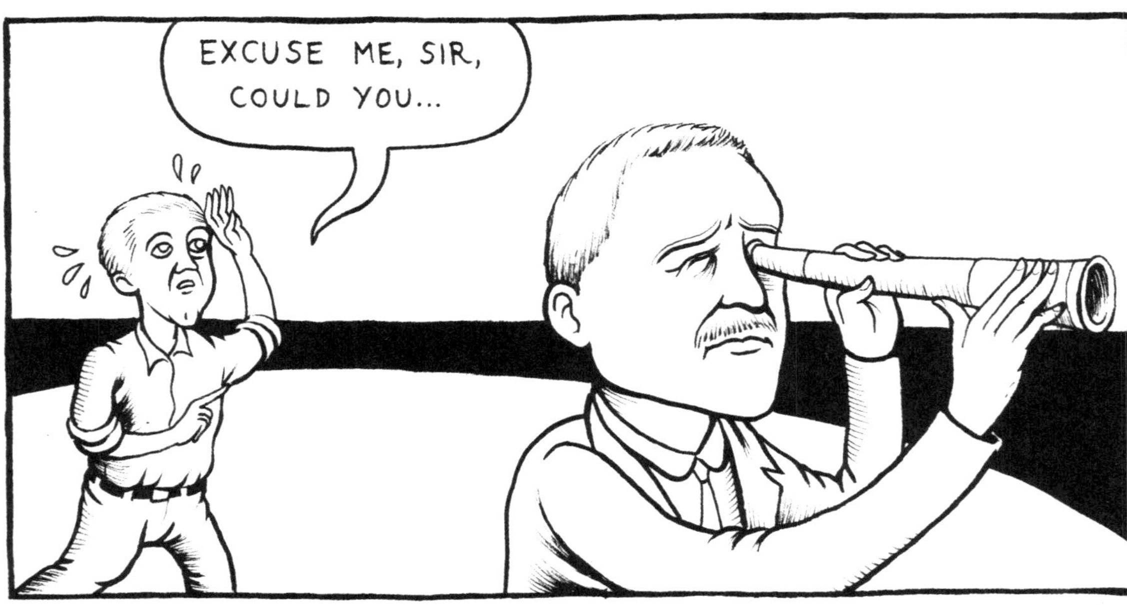
EXCUSE ME, SIR, COULD YOU...

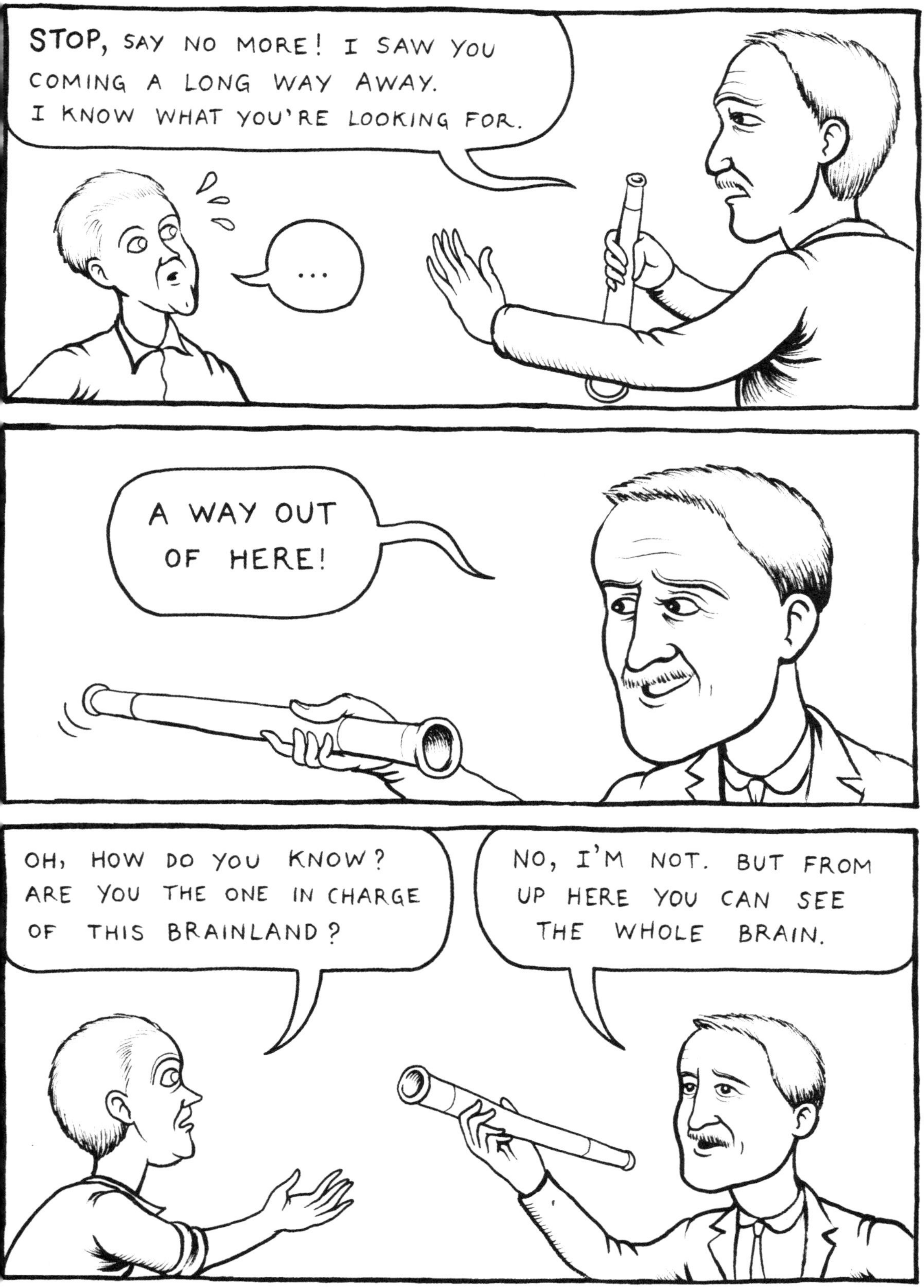
STOP, SAY NO MORE! I SAW YOU COMING A LONG WAY AWAY. I KNOW WHAT YOU'RE LOOKING FOR.
...
A WAY OUT OF HERE!
OH, HOW DO YOU KNOW? ARE YOU THE ONE IN CHARGE OF THIS BRAINLAND?
NO, I'M NOT. BUT FROM UP HERE YOU CAN SEE THE WHOLE BRAIN.

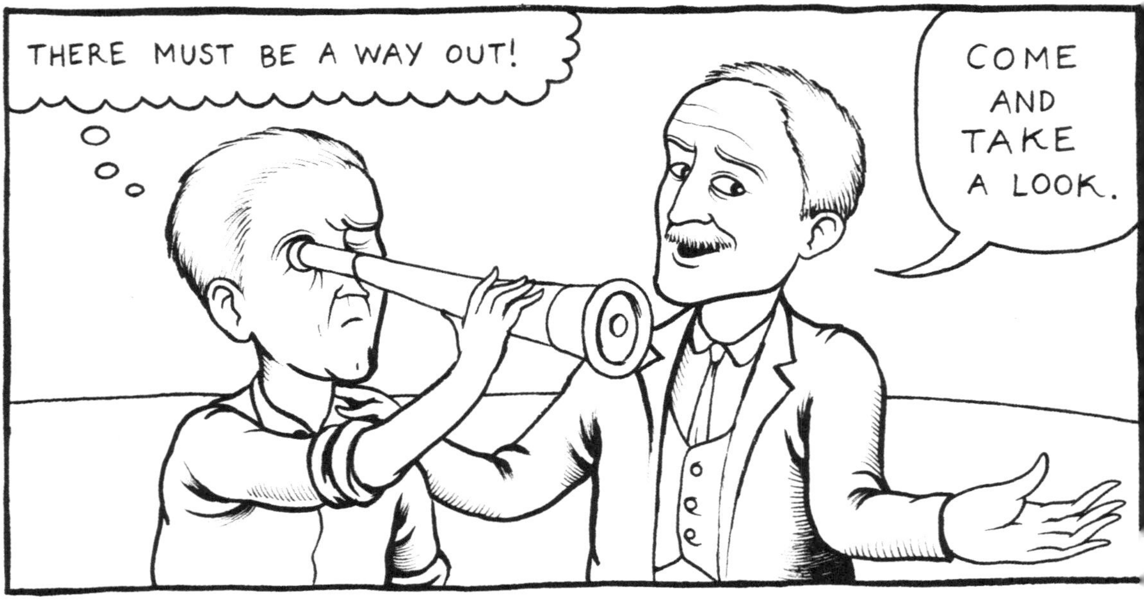
THERE MUST BE A WAY OUT!
COME AND TAKE A LOOK.

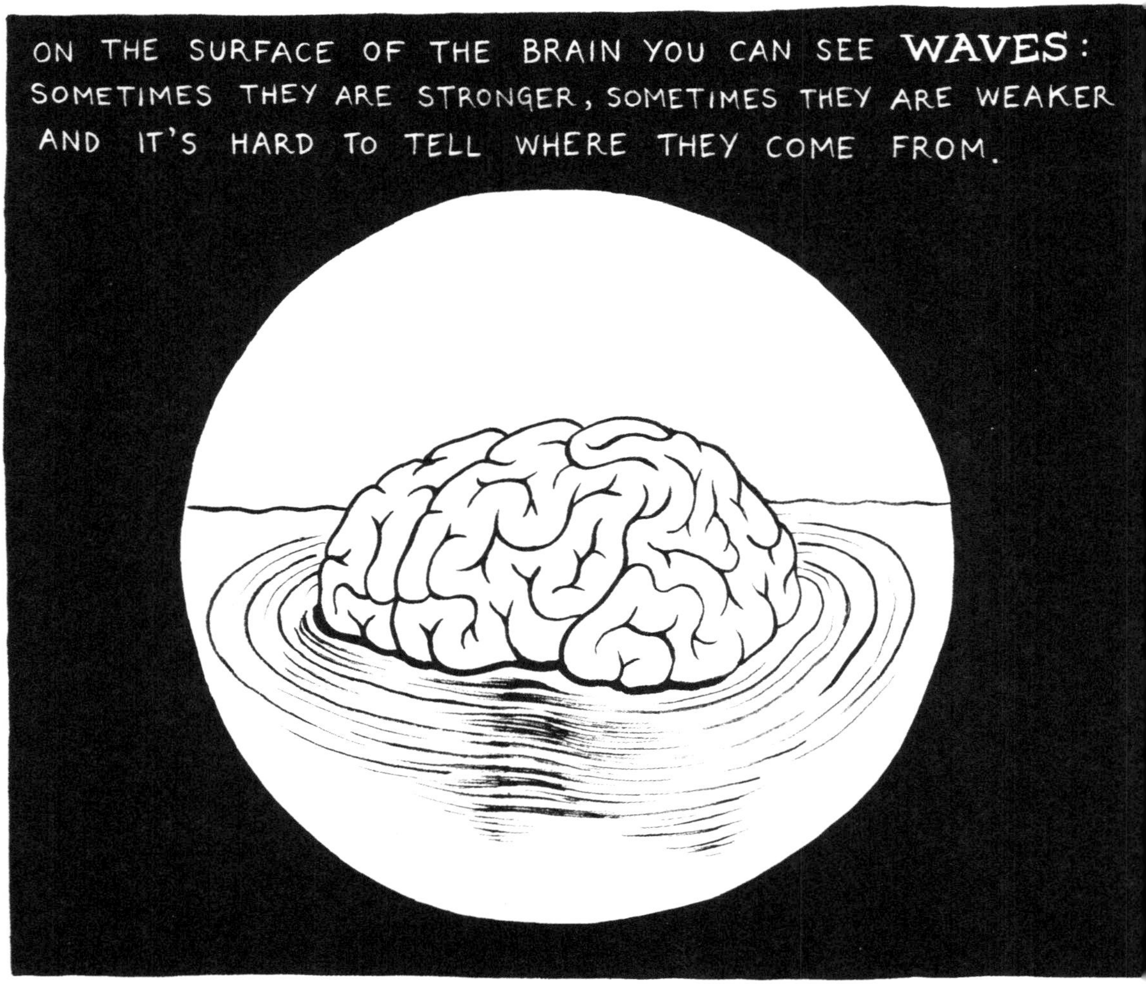
ON THE SURFACE OF THE BRAIN YOU CAN SEE WAVES: SOMETIMES THEY ARE STRONGER, SOMETIMES THEY ARE WEAKER AND IT'S HARD TO TELL WHERE THEY COME FROM.

MY NAME IS **HANS BERGER** AND I WAS THE FIRST TO OBSERVE THESE "BRAIN WAVES" WITH A MACHINE I INVENTED IN 1924 (CALLED AN **ELECTROENCEPHALOGRAM**) WHICH RECORDS THE ELECTRICAL ACTIVITY OF THE BRAIN THROUGH ELECTRODES PLACED ON THE SCALP.

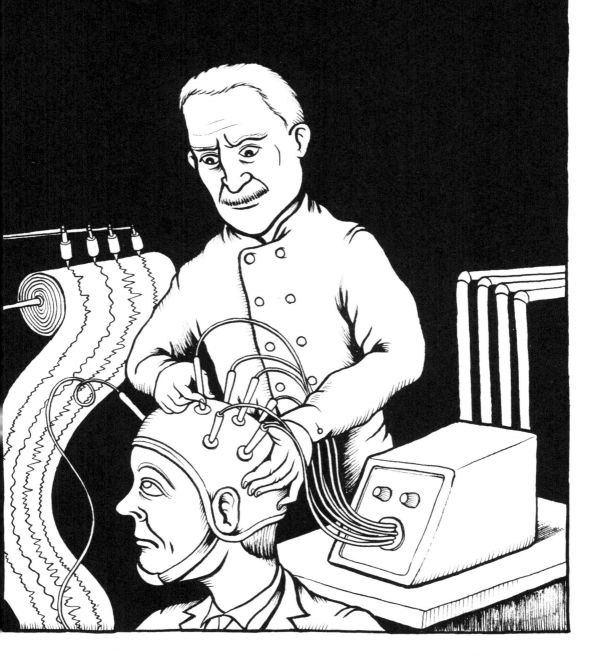

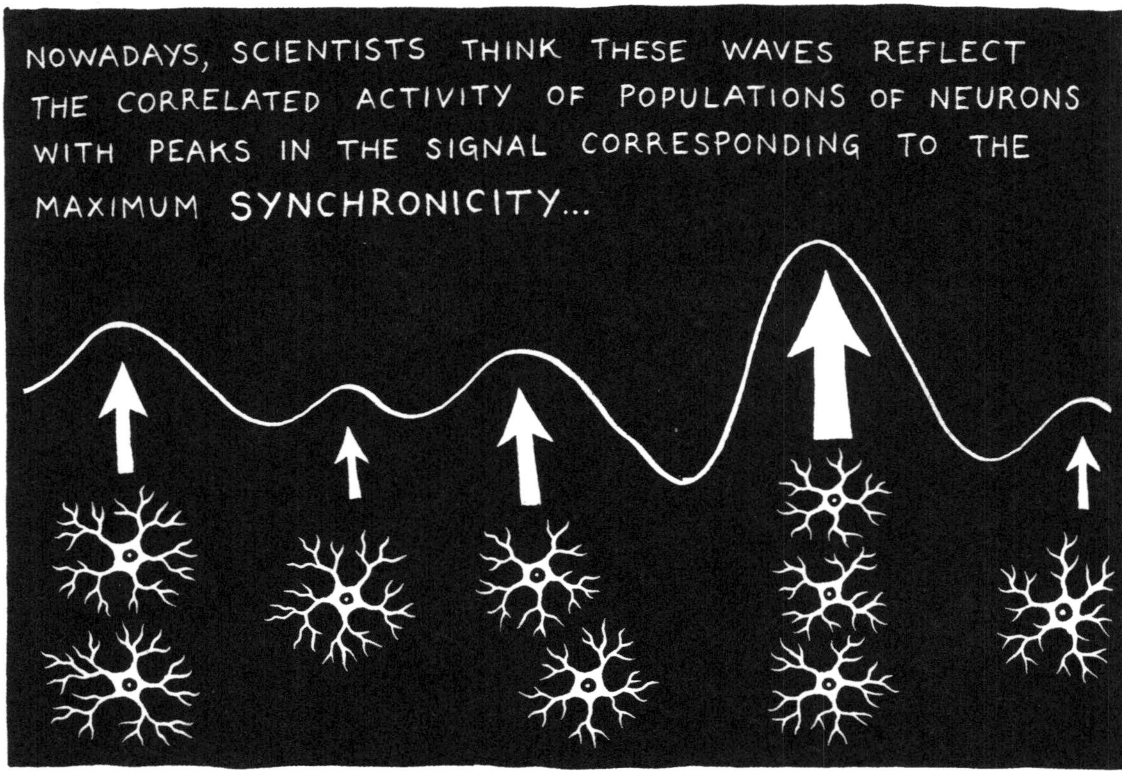
NOWADAYS, SCIENTISTS THINK THESE WAVES REFLECT THE CORRELATED ACTIVITY OF POPULATIONS OF NEURONS WITH PEAKS IN THE SIGNAL CORRESPONDING TO THE MAXIMUM SYNCHRONICITY...

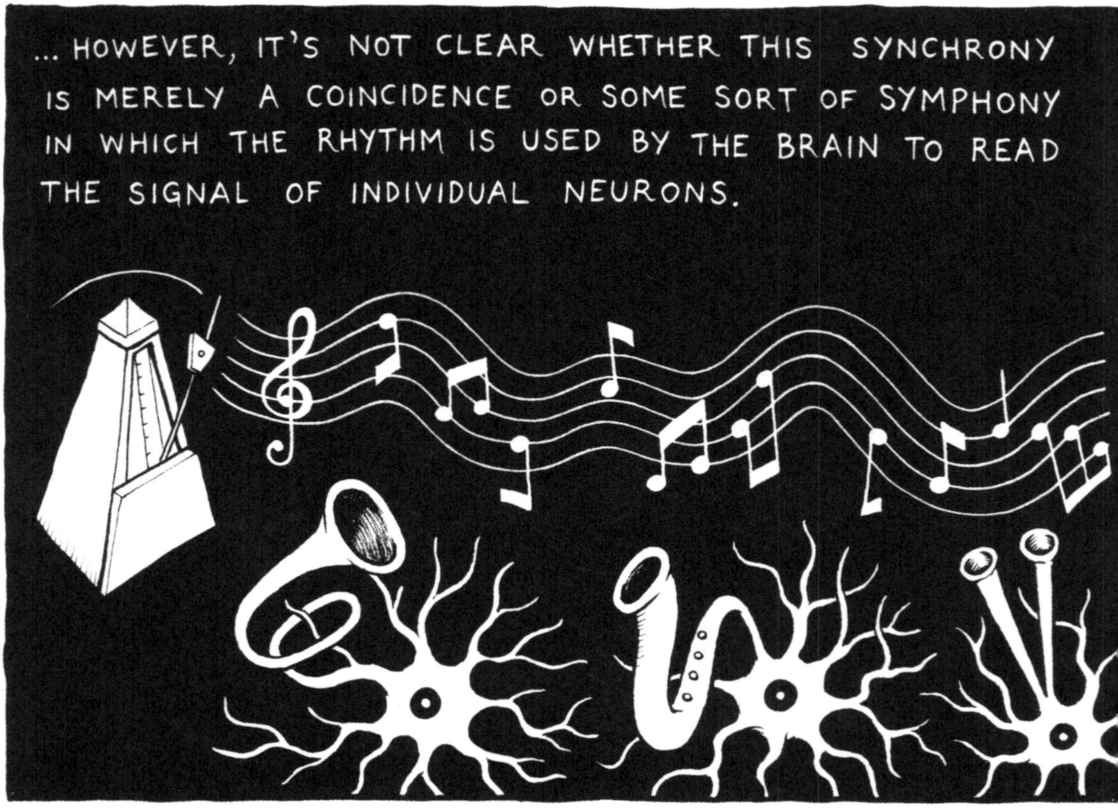
... HOWEVER, IT'S NOT CLEAR WHETHER THIS SYNCHRONY IS MERELY A COINCIDENCE OR SOME SORT OF SYMPHONY IN WHICH THE RHYTHM IS USED BY THE BRAIN TO READ THE SIGNAL OF INDIVIDUAL NEURONS.

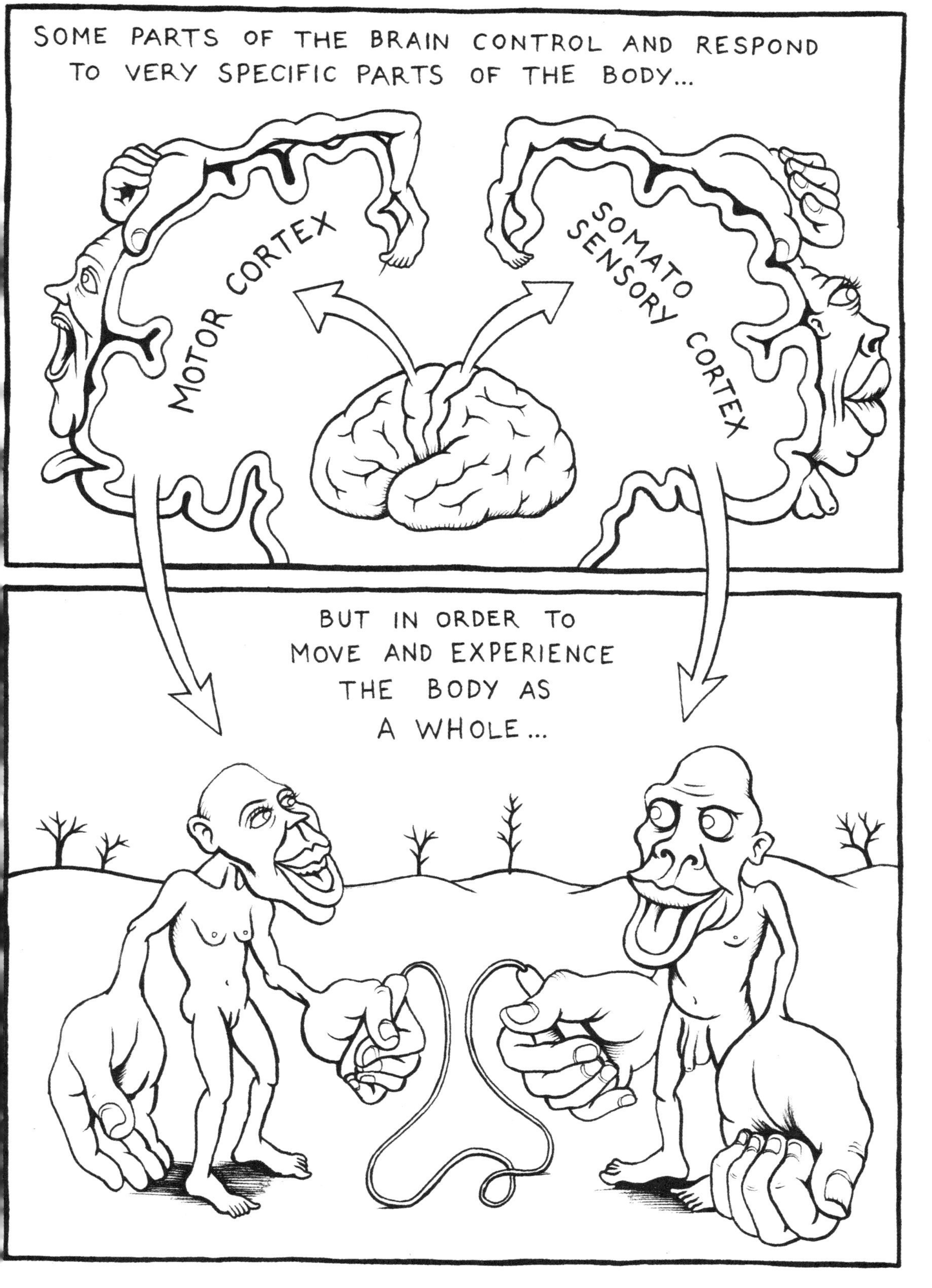
SOME PARTS OF THE BRAIN CONTROL AND RESPOND TO VERY SPECIFIC PARTS OF THE BODY...
MOTOR CORTEX
SOMATO SENSORY CORTEX
BUT IN ORDER TO MOVE AND EXPERIENCE THE BODY AS A WHOLE...

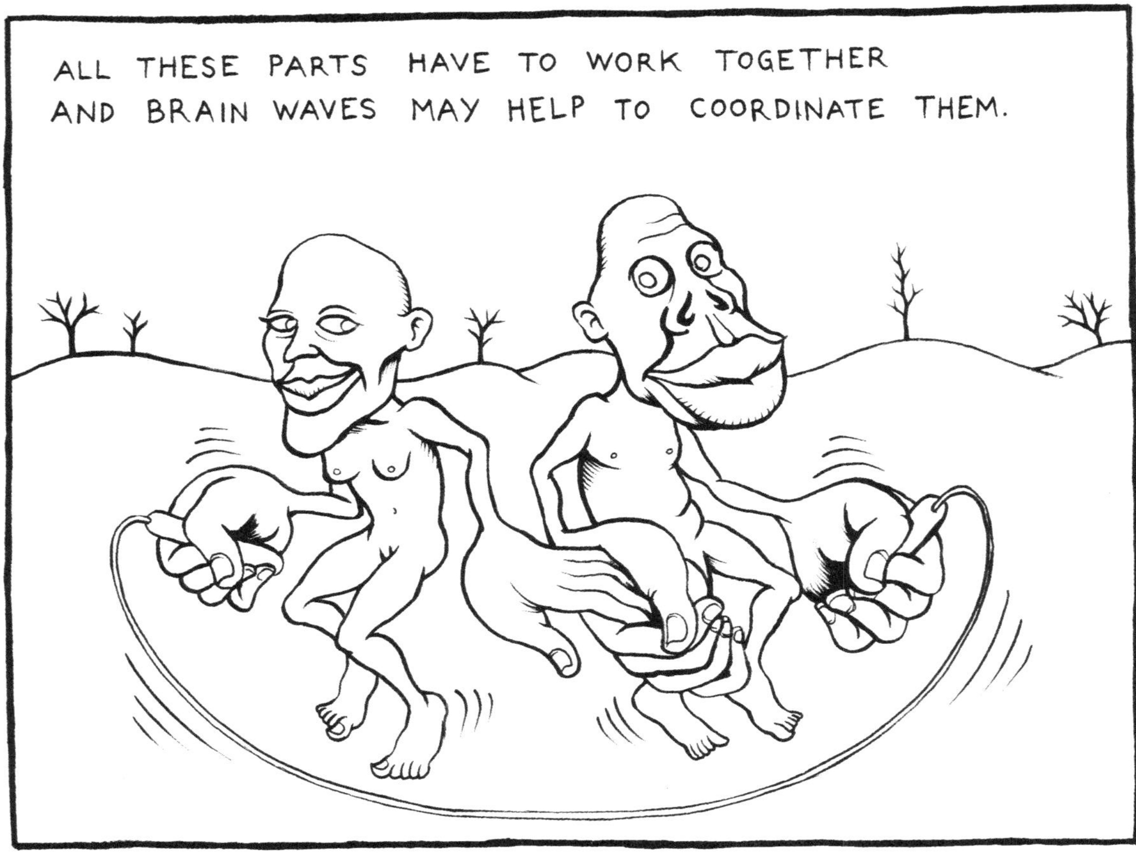
ALL THESE PARTS HAVE TO WORK TOGETHER AND BRAIN WAVES MAY HELP TO COORDINATE THEM.

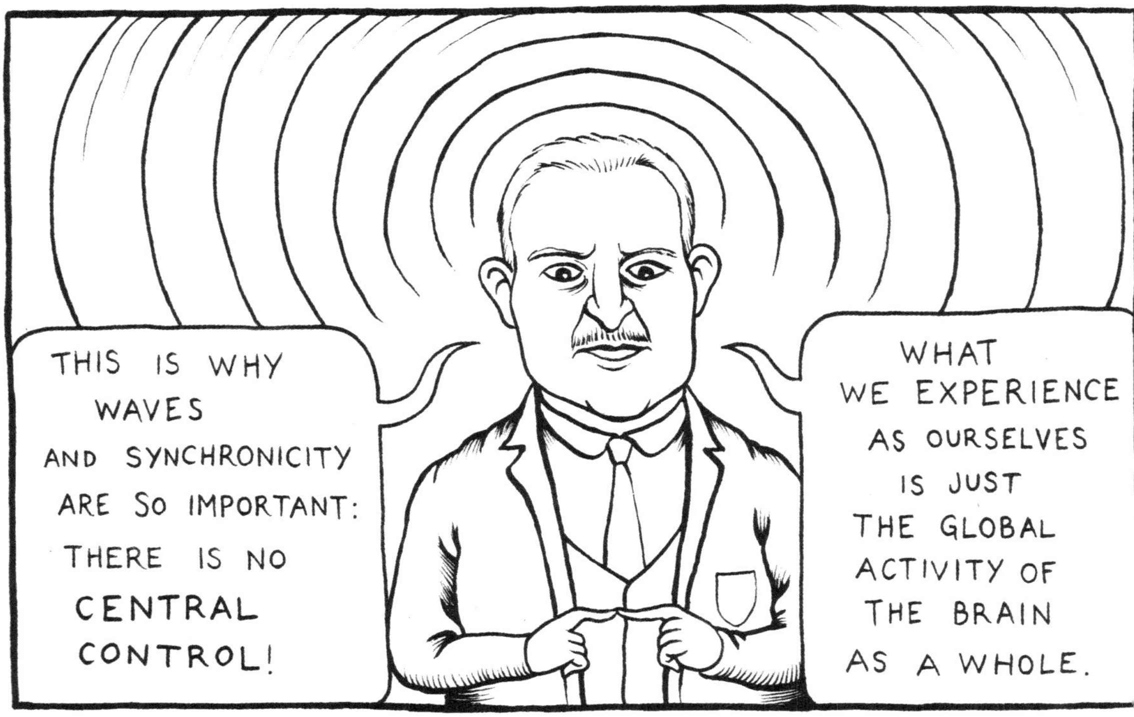
THIS IS WHY WAVES AND SYNCHRONICITY ARE SO IMPORTANT: THERE IS NO CENTRAL CONTROL!
WHAT WE EXPERIENCE AS OURSELVES IS JUST THE GLOBAL ACTIVITY OF THE BRAIN AS A WHOLE.

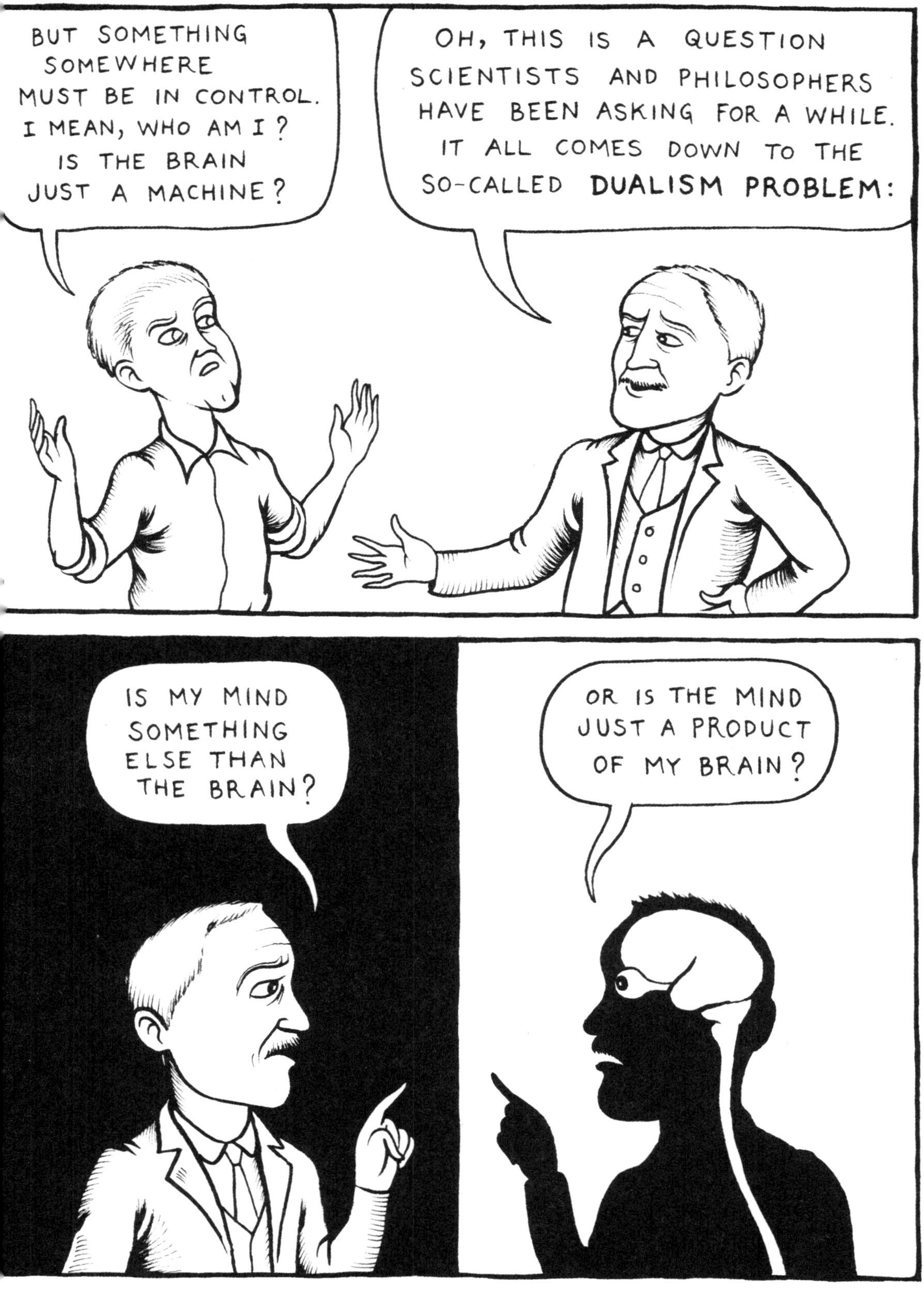
BUT SOMETHING SOMEWHERE MUST BE IN CONTROL. I MEAN, WHO AM I? IS THE BRAIN JUST A MACHINE?
OH, THIS IS A QUESTION SCIENTISTS AND PHILOSOPHERS HAVE BEEN ASKING FOR A WHILE. IT ALL COMES DOWN TO THE SO-CALLED DUALISM PROBLEM:
IS MY MIND SOMETHING ELSE THAN THE BRAIN?
OR IS THE MIND JUST A PRODUCT OF MY BRAIN?

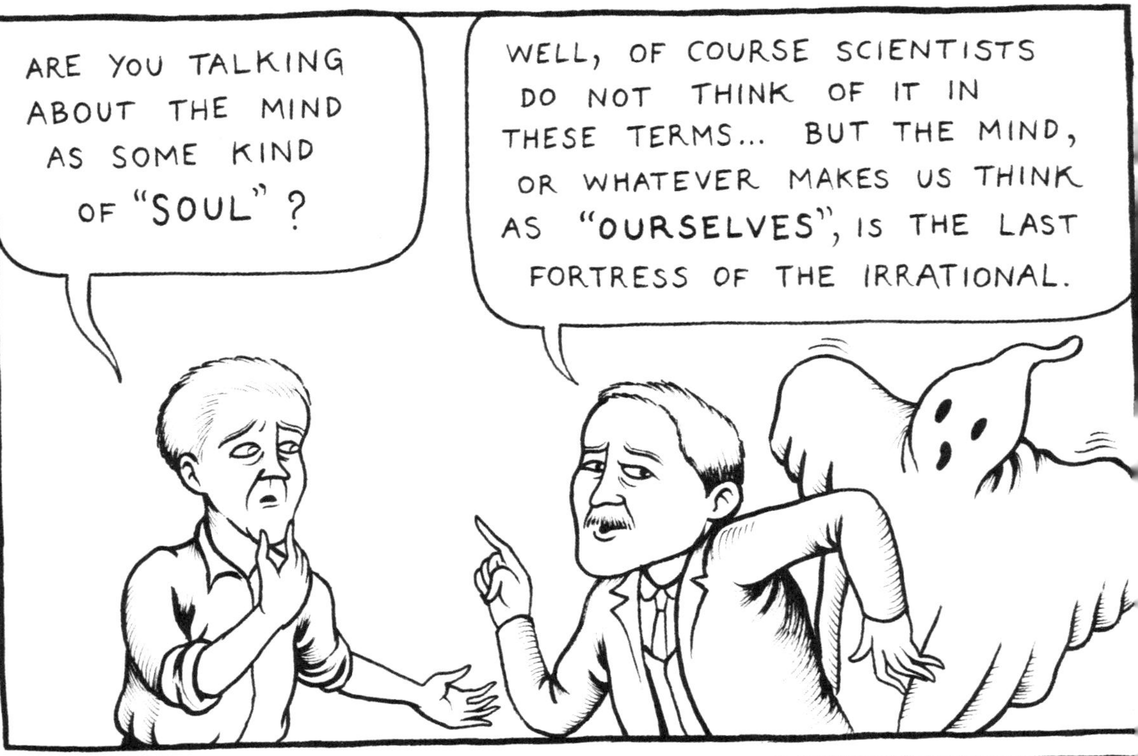
ARE YOU TALKING ABOUT THE MIND AS SOME KIND OF "SOUL"?
WELL, OF COURSE SCIENTISTS DO NOT THINK OF IT IN THESE TERMS... BUT THE MIND, OR WHATEVER MAKES US THINK AS "OURSELVES", IS THE LAST FORTRESS OF THE IRRATIONAL.

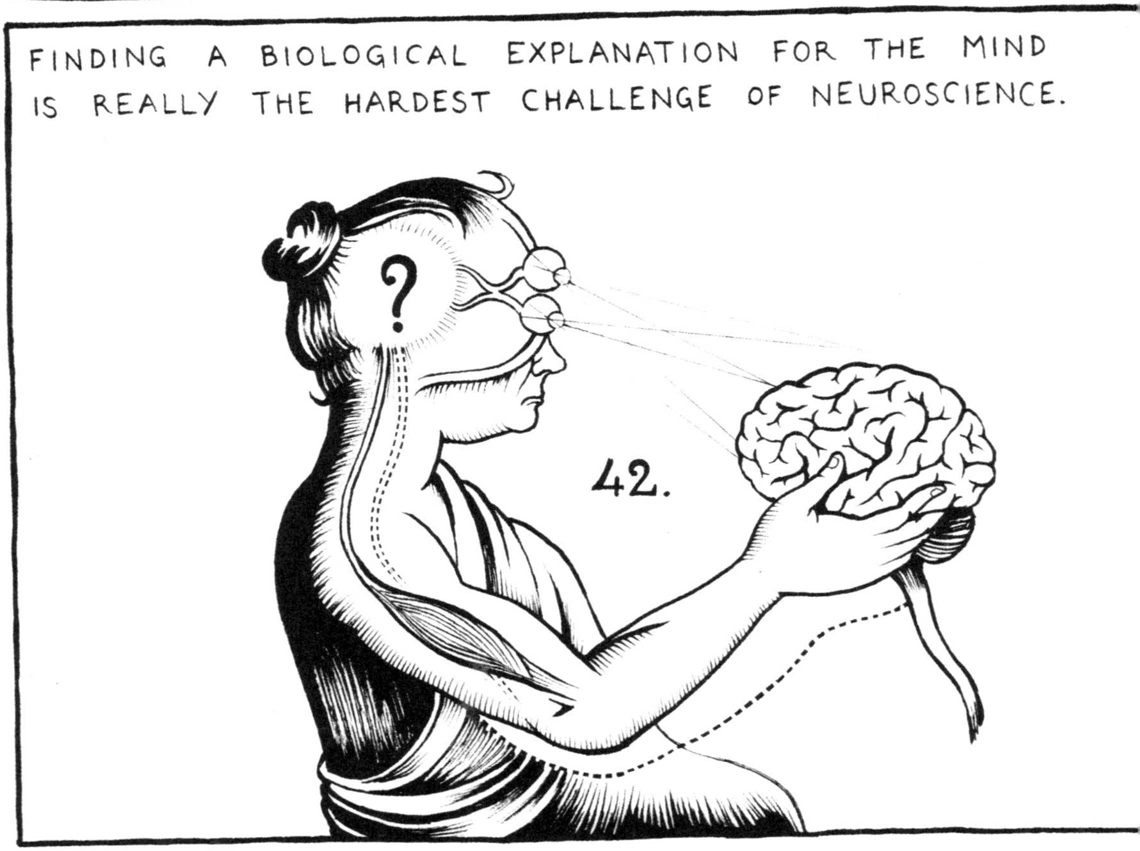
FINDING A BIOLOGICAL EXPLANATION FOR THE MIND IS REALLY THE HARDEST CHALLENGE OF NEUROSCIENCE.
?
42.

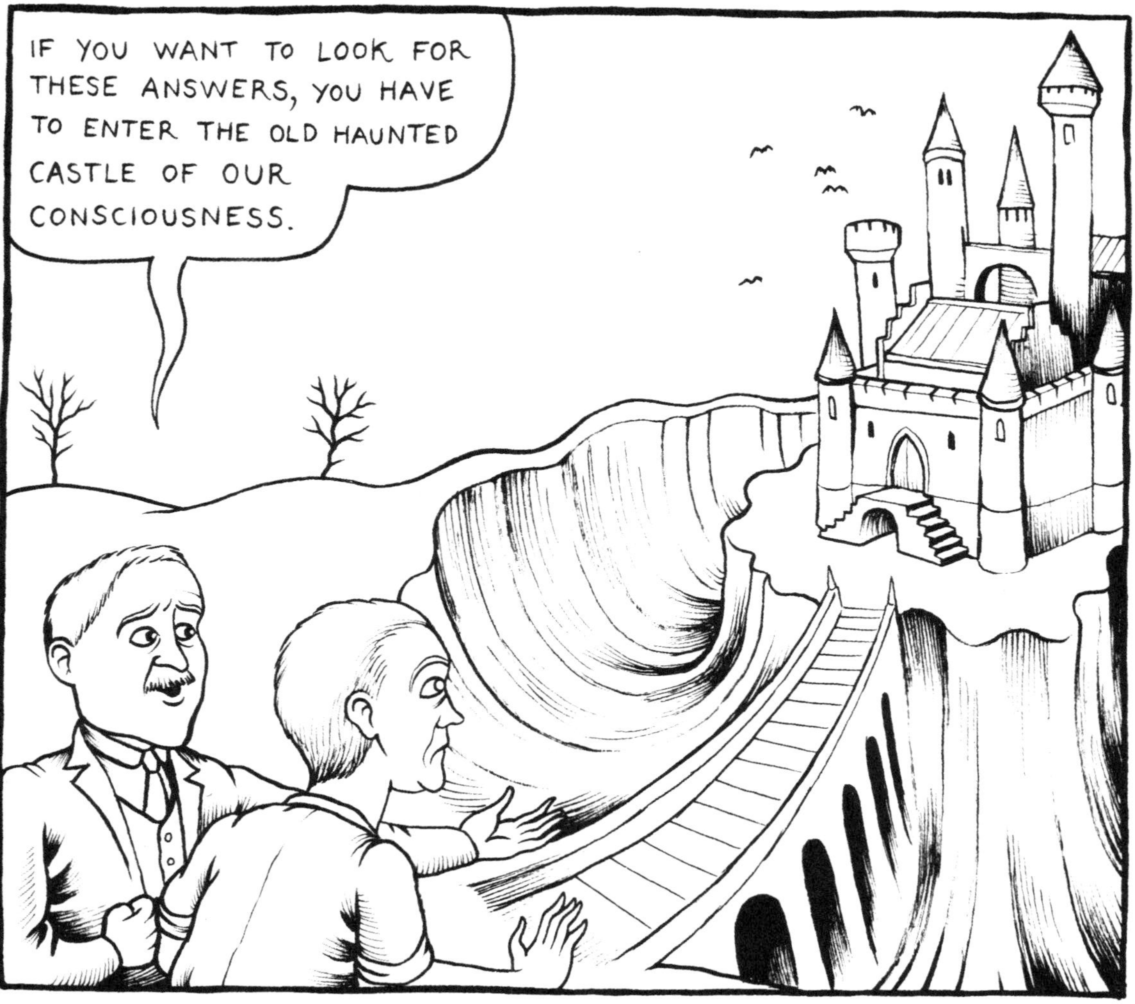
IF YOU WANT TO LOOK FOR THESE ANSWERS, YOU HAVE TO ENTER THE OLD HAUNTED CASTLE OF OUR CONSCIOUSNESS.

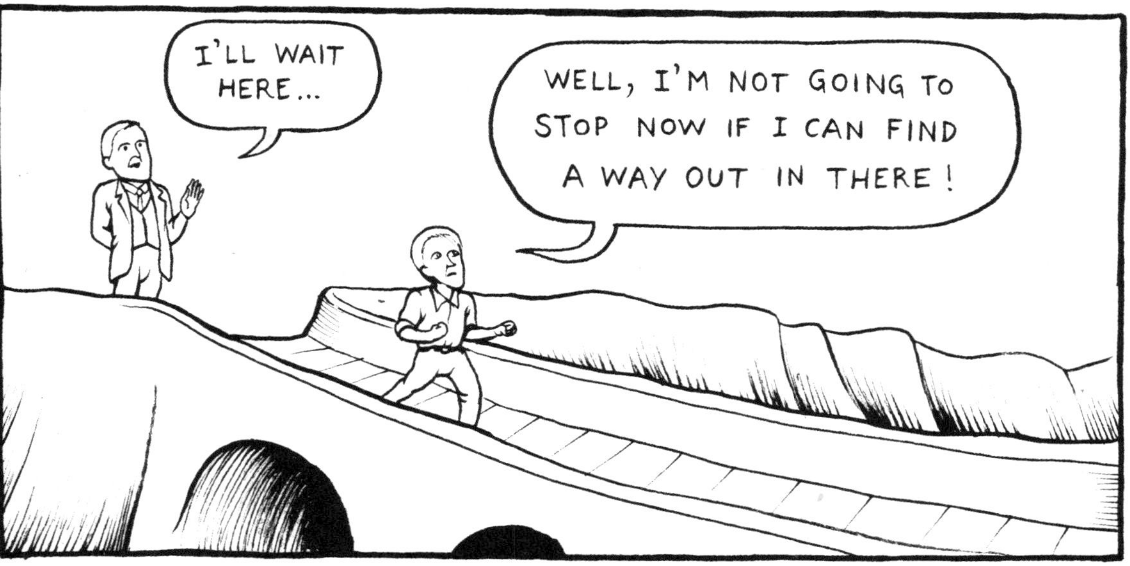
I'LL WAIT HERE...
WELL, I'M NOT GOING TO STOP NOW IF I CAN FIND A WAY OUT IN THERE!

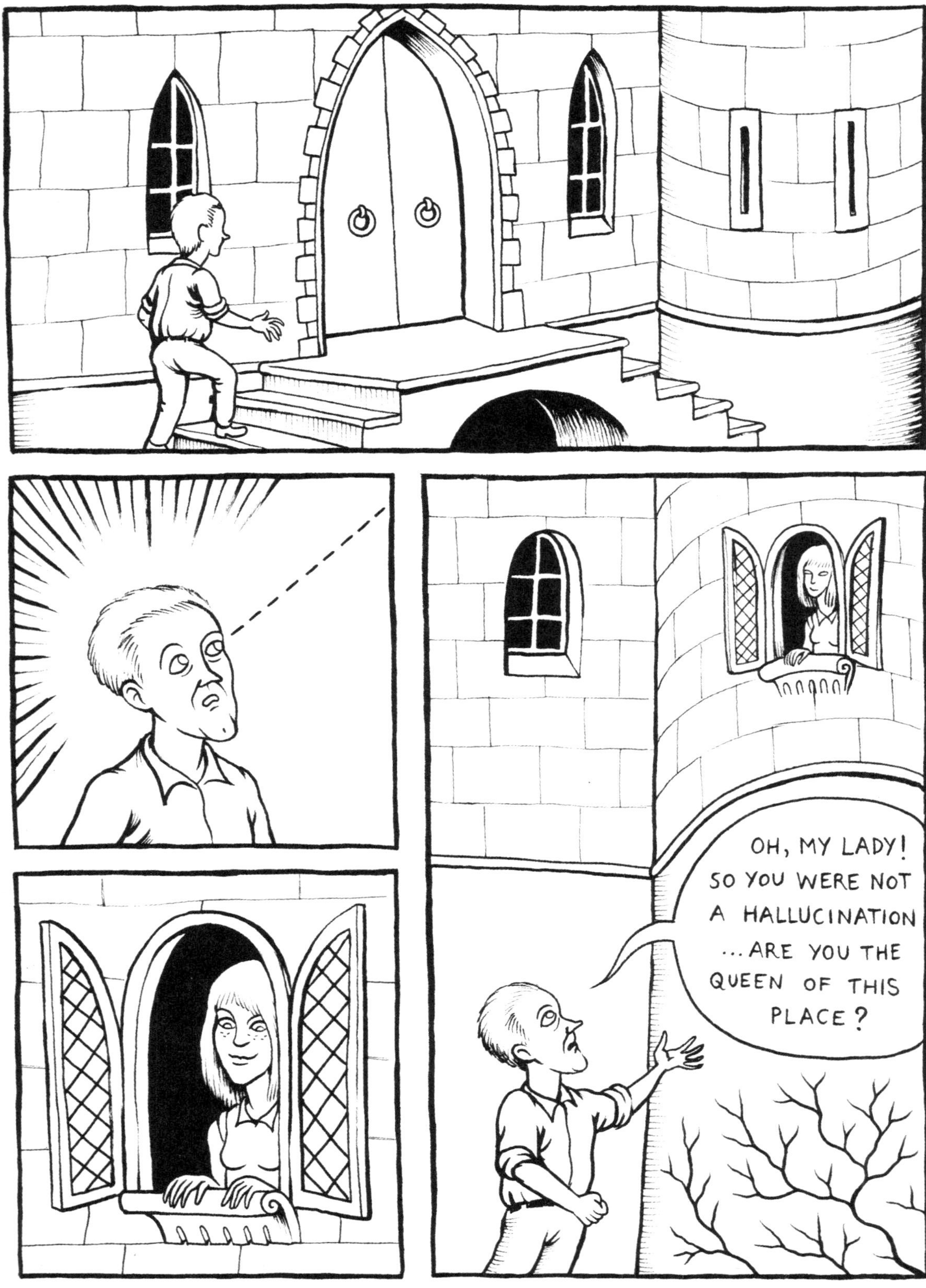
OH, MY LADY! SO YOU WERE NOT A HALLUCINATION ...ARE YOU THE QUEEN OF THIS PLACE?

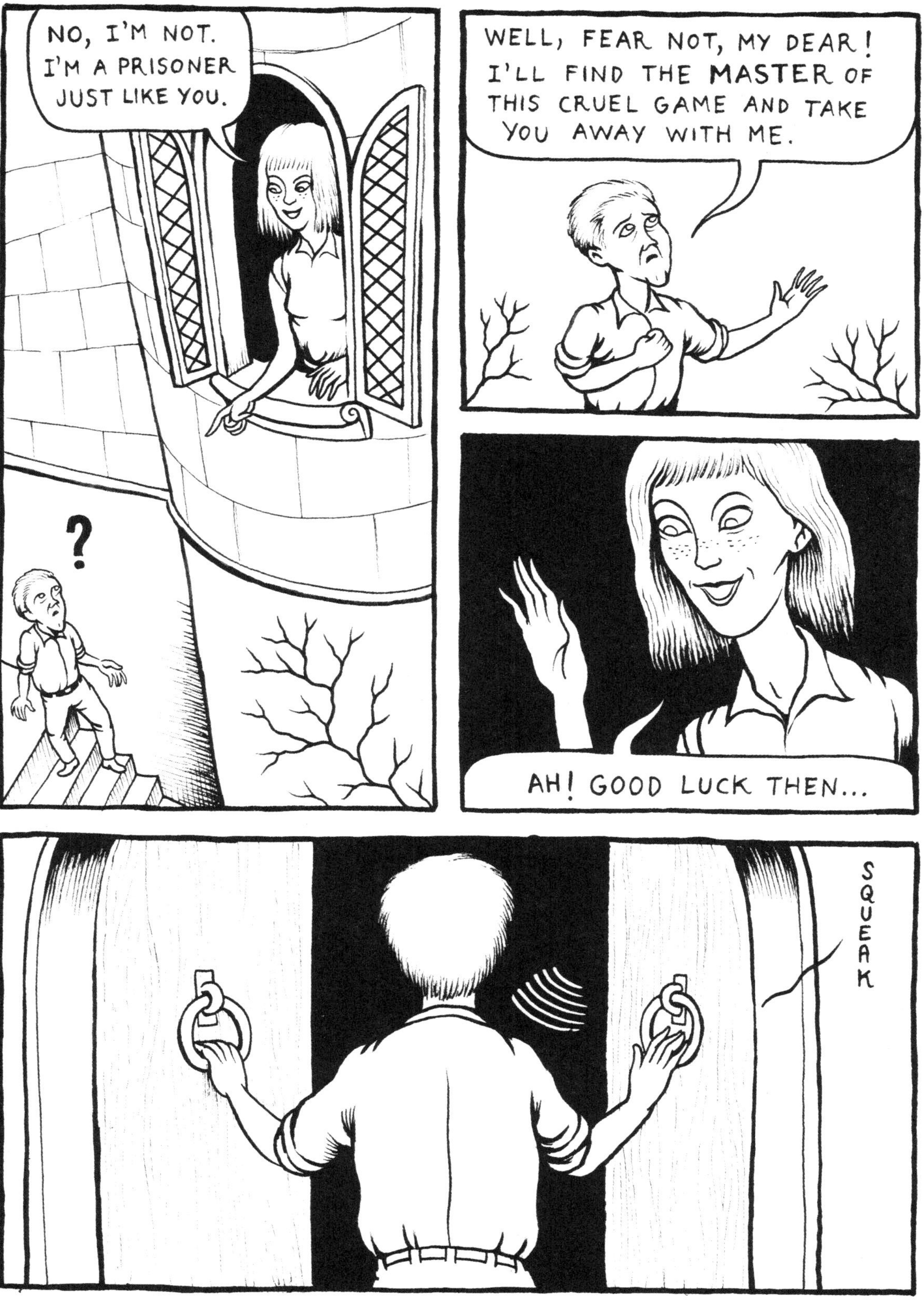
NO, I'M NOT. I'M A PRISONER JUST LIKE YOU.
WELL, FEAR NOT, MY DEAR! I'LL FIND THE MASTER OF THIS CRUEL GAME AND TAKE YOU AWAY WITH ME.
?
AH! GOOD LUCK THEN...
SQUEAK

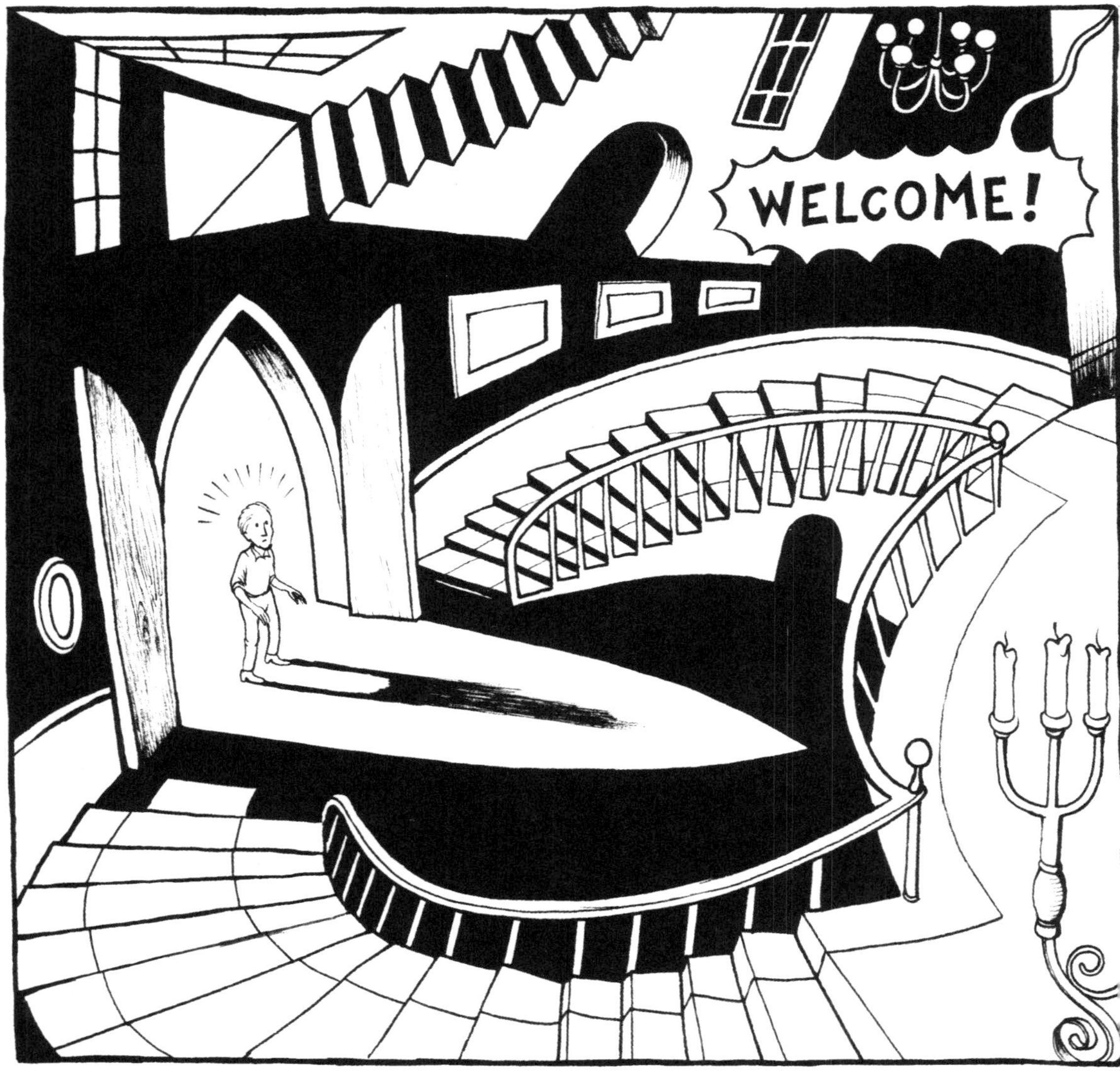
WELCOME!

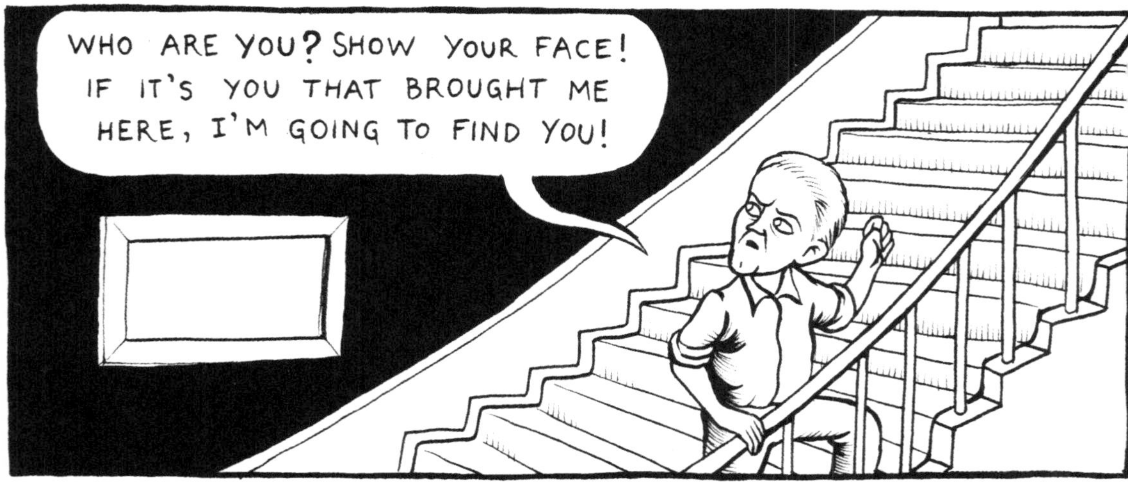
WHO ARE YOU? SHOW YOUR FACE! IF IT'S YOU THAT BROUGHT ME HERE, I'M GOING TO FIND YOU!

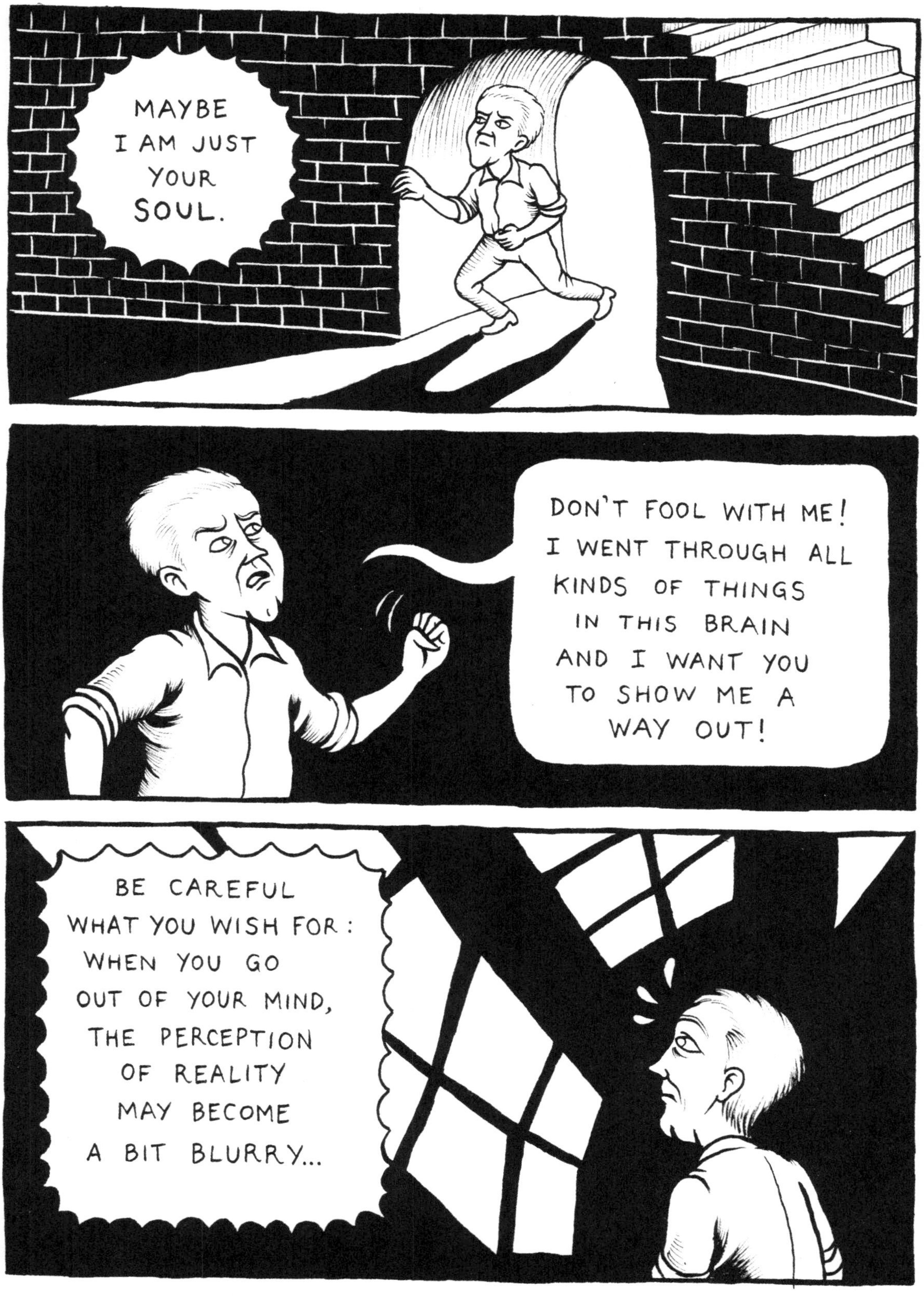
MAYBE I AM JUST YOUR SOUL.
DON'T FOOL WITH ME! I WENT THROUGH ALL KINDS OF THINGS IN THIS BRAIN AND I WANT YOU TO SHOW ME A WAY OUT!
BE CAREFUL WHAT YOU WISH FOR: WHEN YOU GO OUT OF YOUR MIND, THE PERCEPTION OF REALITY MAY BECOME A BIT BLURRY...

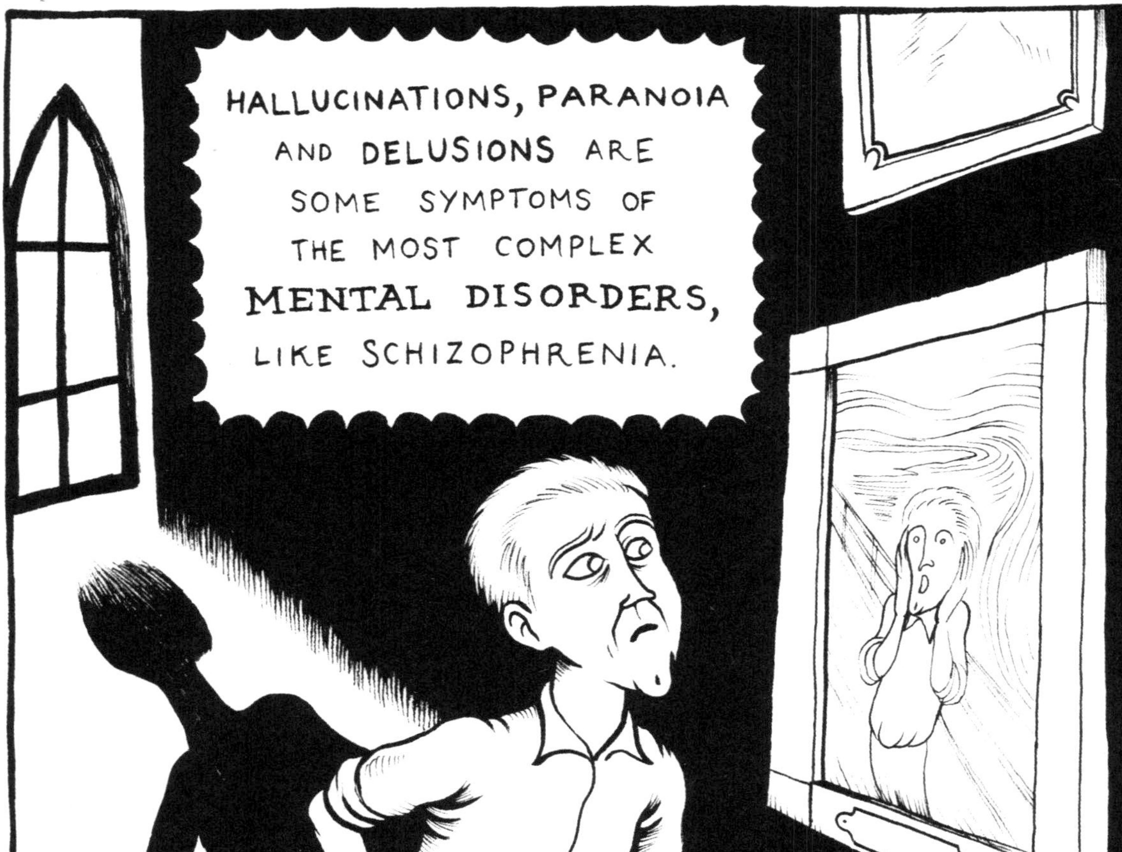
HALLUCINATIONS, PARANOIA AND DELUSIONS ARE SOME SYMPTOMS OF THE MOST COMPLEX MENTAL DISORDERS, LIKE SCHIZOPHRENIA.

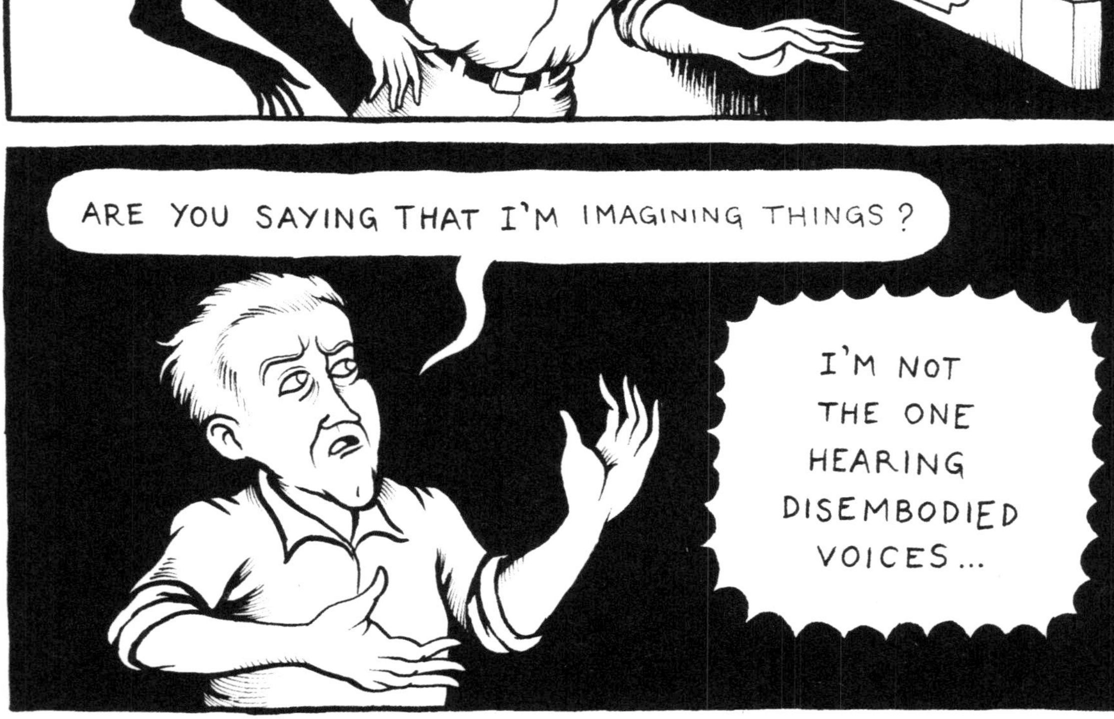
ARE YOU SAYING THAT I'M IMAGINING THINGS?
I'M NOT THE ONE HEARING DISEMBODIED VOICES...

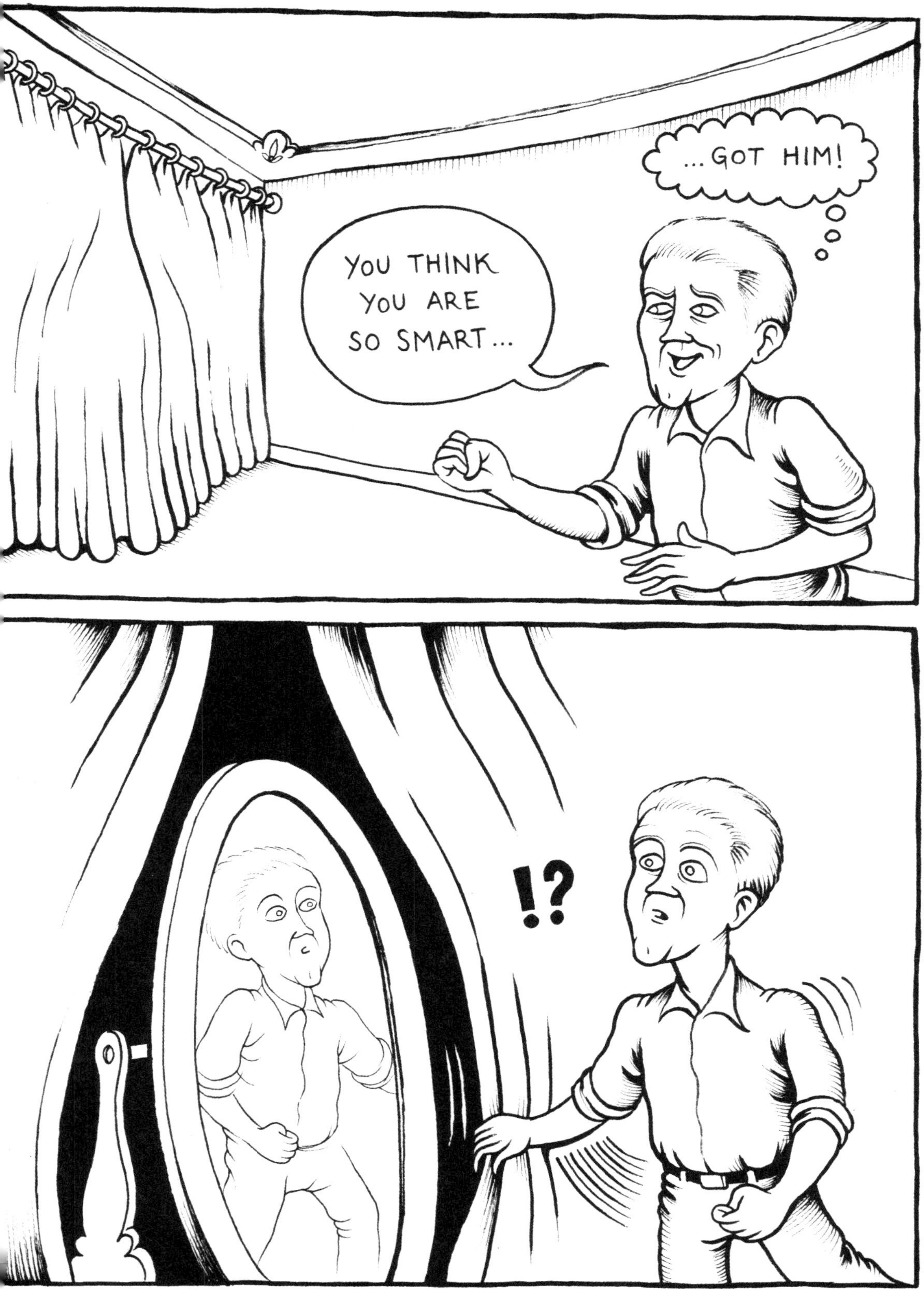
...GOT HIM!
YOU THINK YOU ARE SO SMART...
!?

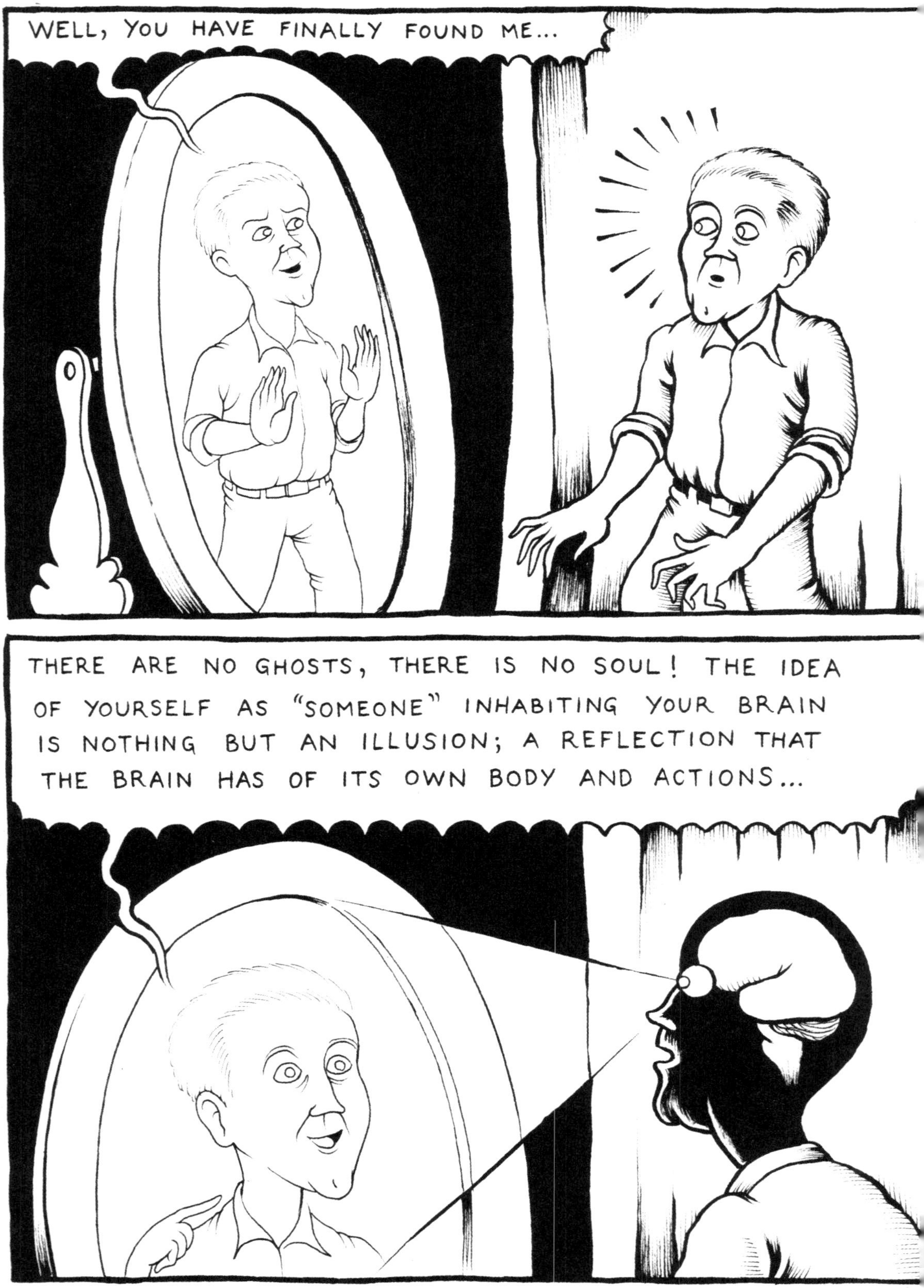
WELL, YOU HAVE FINALLY FOUND ME...
THERE ARE NO GHOSTS, THERE IS NO SOUL! THE IDEA OF YOURSELF AS "SOMEONE" INHABITING YOUR BRAIN IS NOTHING BUT AN ILLUSION; A REFLECTION THAT THE BRAIN HAS OF ITS OWN BODY AND ACTIONS...

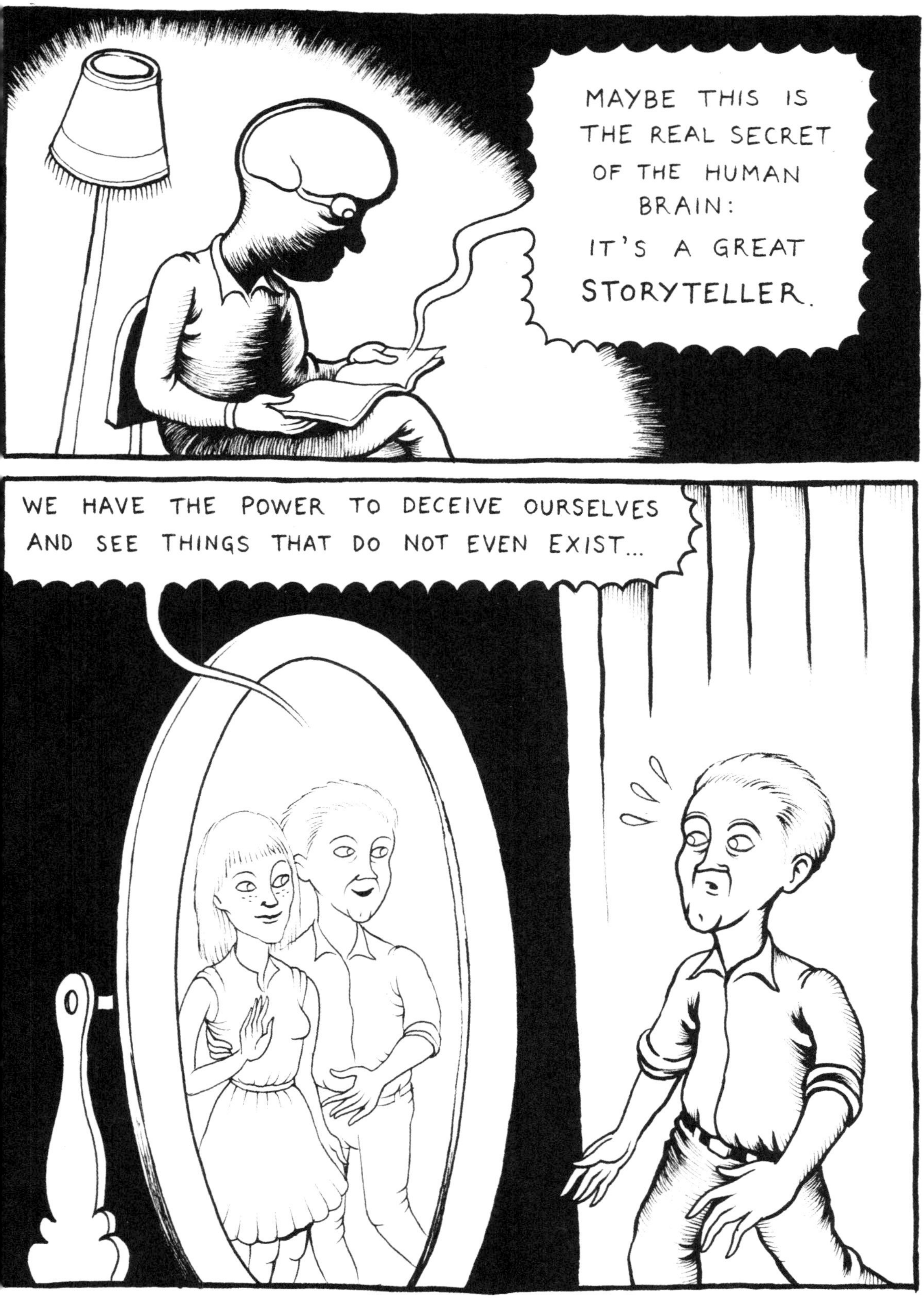
MAYBE THIS IS THE REAL SECRET OF THE HUMAN BRAIN: IT'S A GREAT STORYTELLER.
WE HAVE THE POWER TO DECEIVE OURSELVES AND SEE THINGS THAT DO NOT EVEN EXIST...

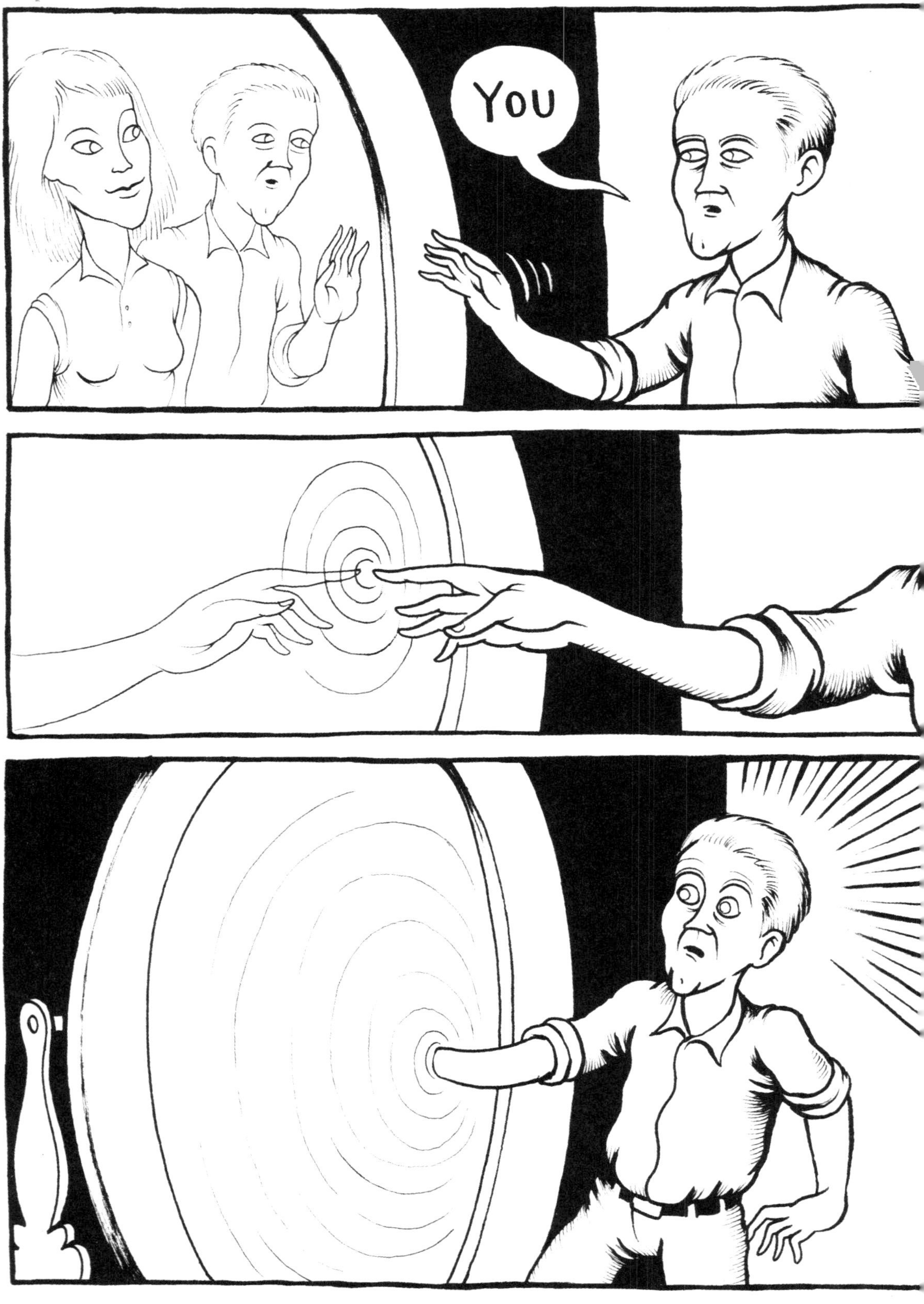
YOU

EPILOGUE

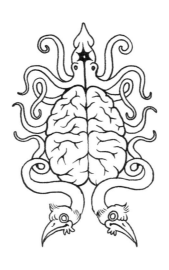

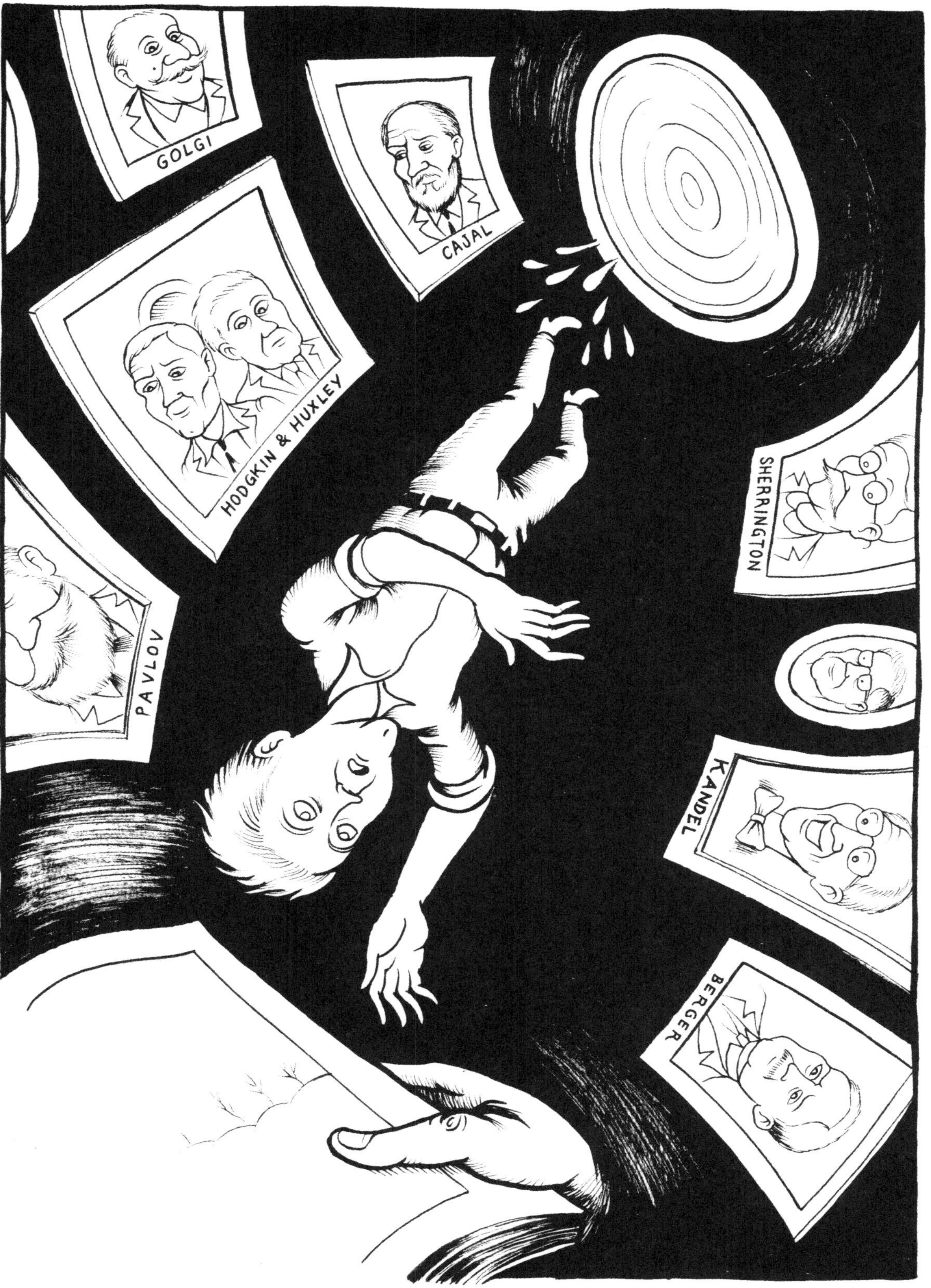
GOLGI
CAJAL
HODGKIN & HUXLEY
PAVLOV
SHERRINGTON
KANDEL
BERGER

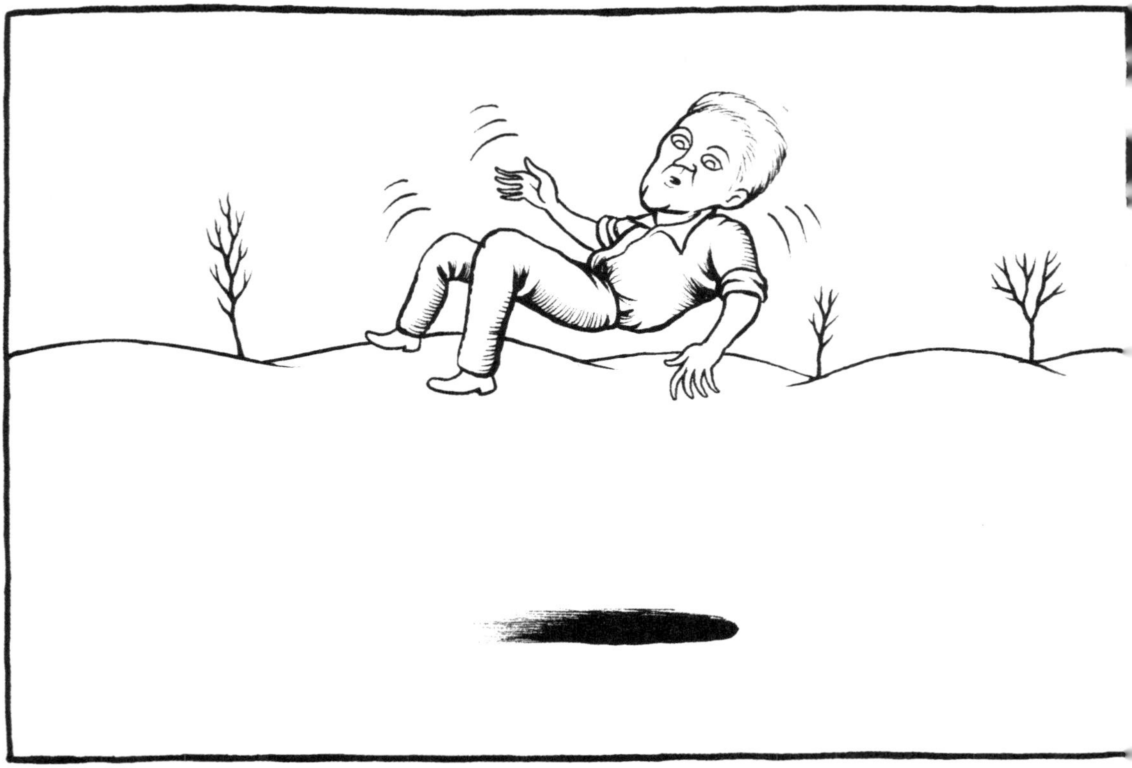

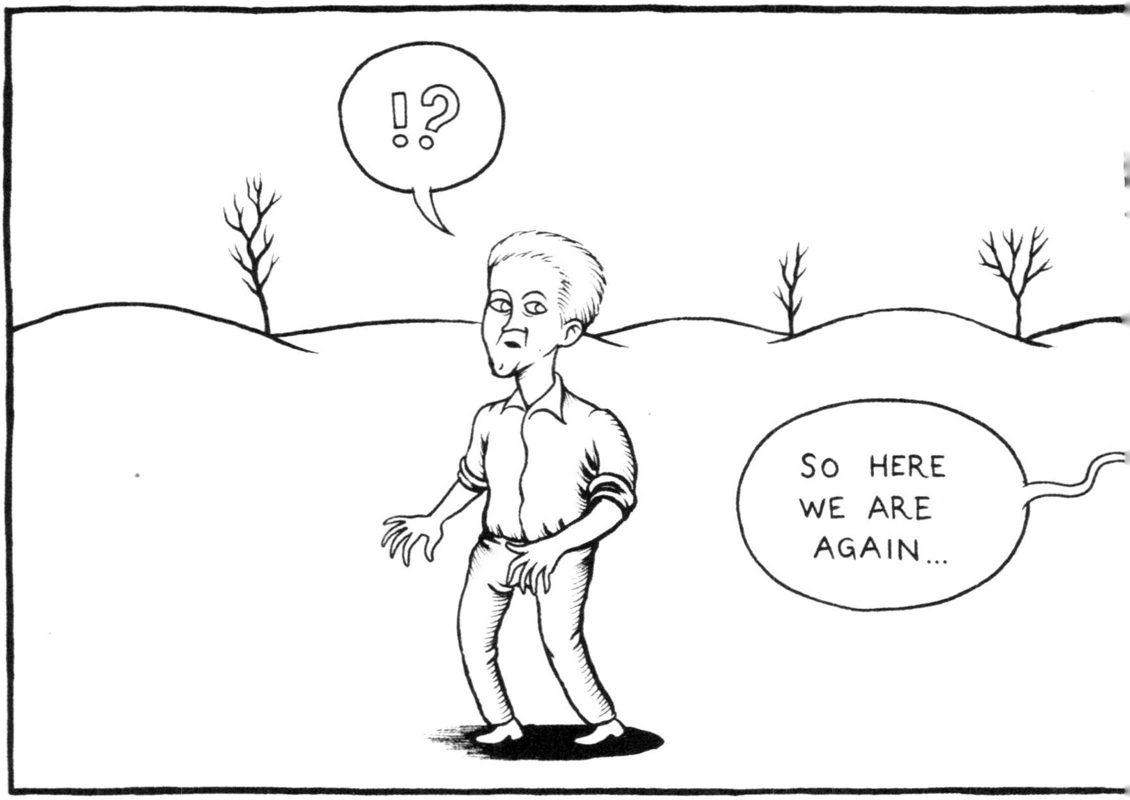
!?
SO HERE WE ARE AGAIN...

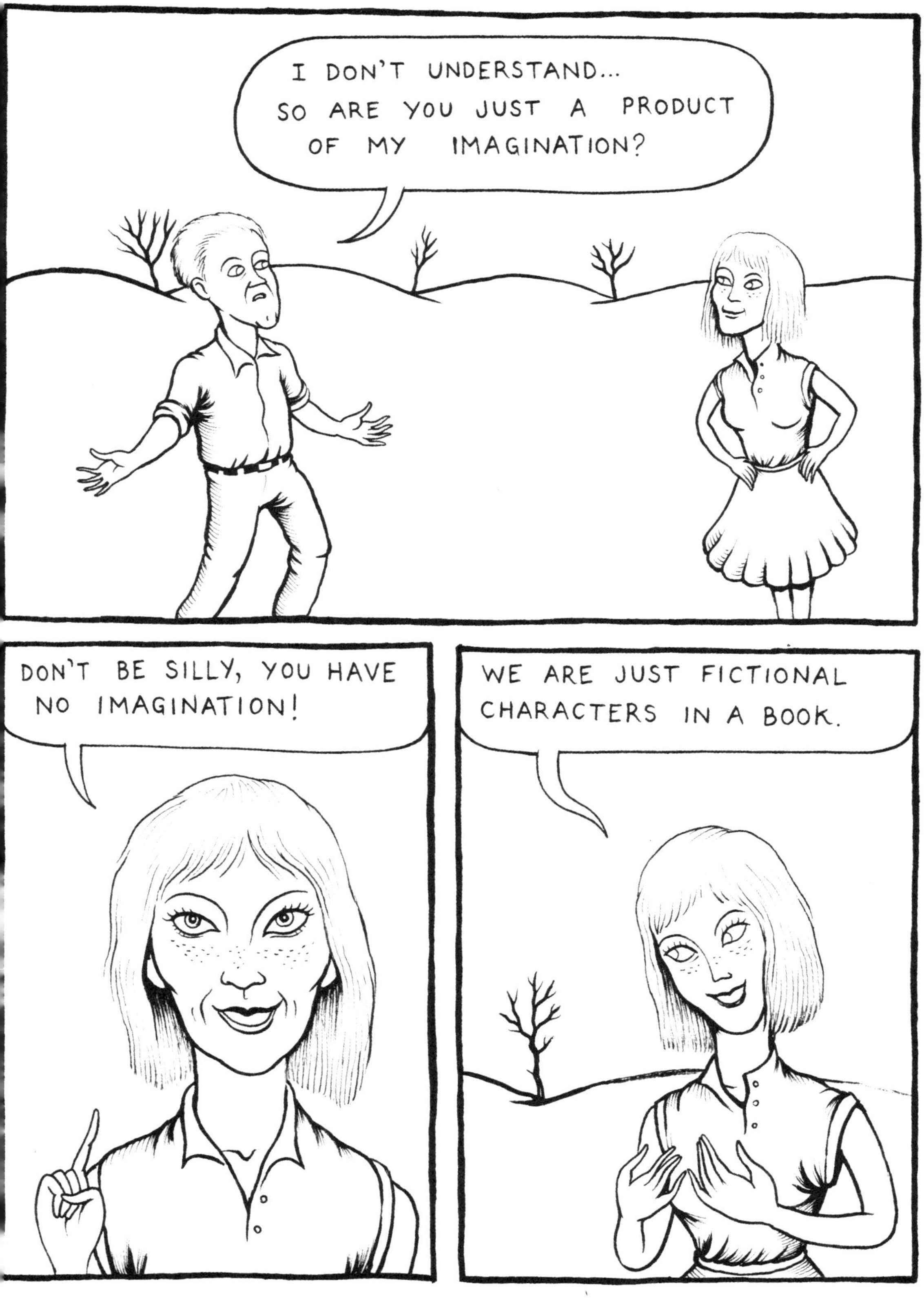

I DON'T UNDERSTAND... SO ARE YOU JUST A PRODUCT OF MY IMAGINATION?
DON'T BE SILLY, YOU HAVE NO IMAGINATION!
WE ARE JUST FICTIONAL CHARACTERS IN A BOOK.

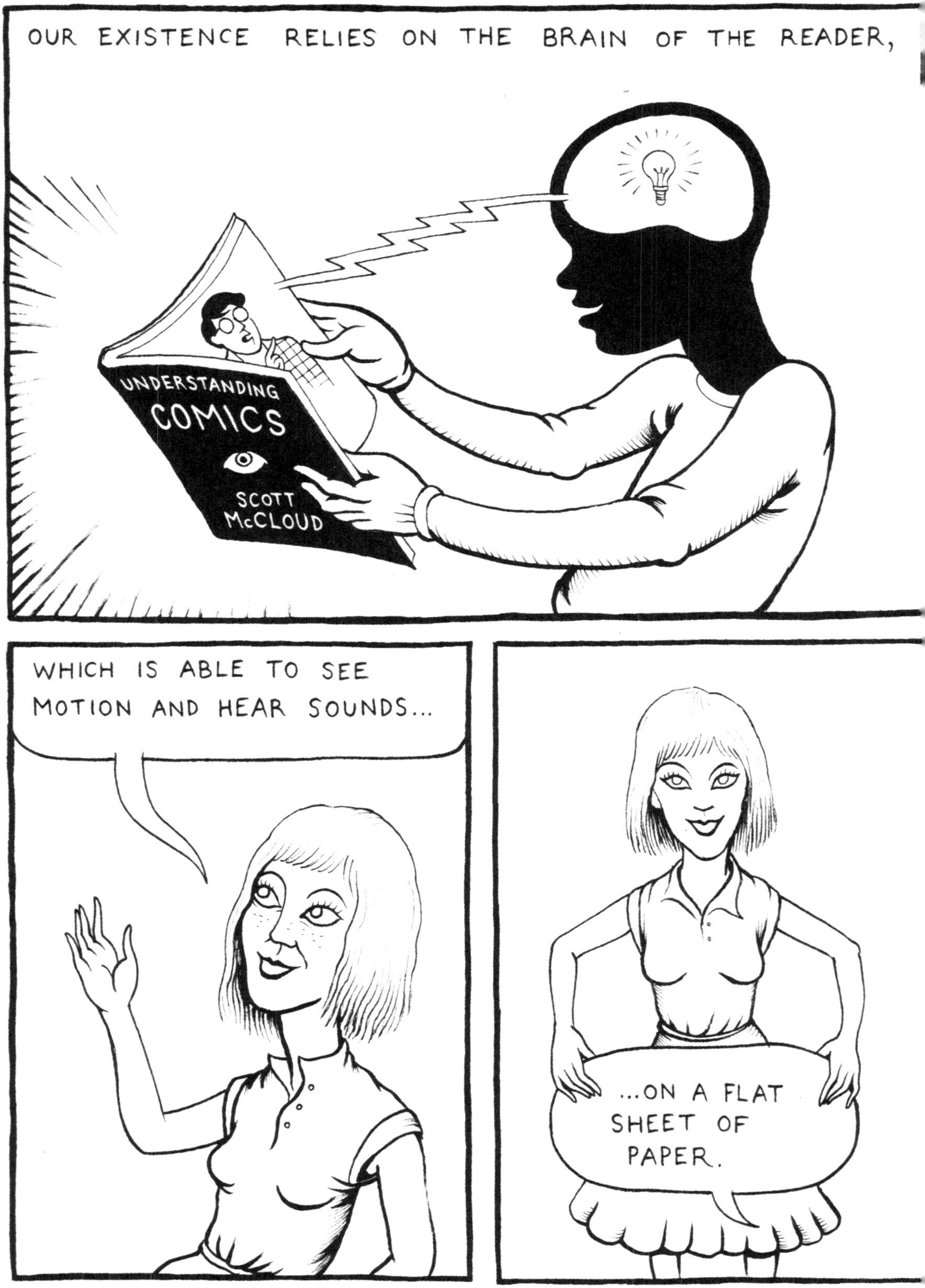
OUR EXISTENCE RELIES ON THE BRAIN OF THE READER,
UNDERSTANDING COMICS
SCOTT McCLOUD
WHICH IS ABLE TO SEE MOTION AND HEAR SOUNDS...
...ON A FLAT SHEET OF PAPER.

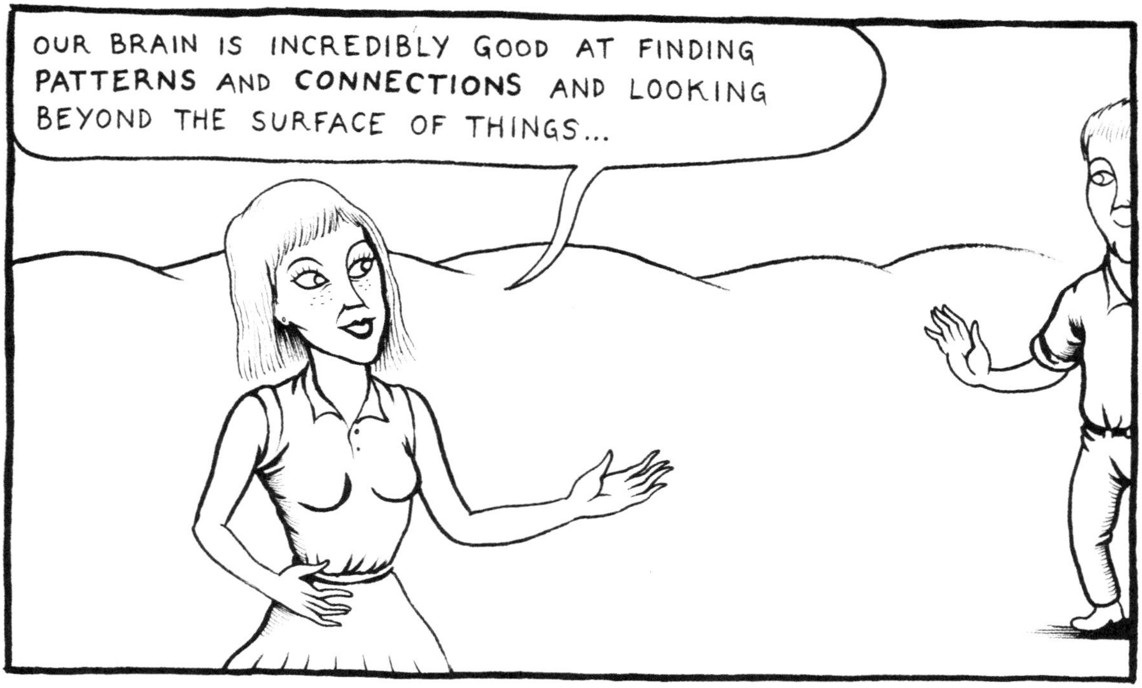
OUR BRAIN IS INCREDIBLY GOOD AT FINDING **PATTERNS** AND **CONNECTIONS** AND LOOKING BEYOND THE SURFACE OF THINGS...

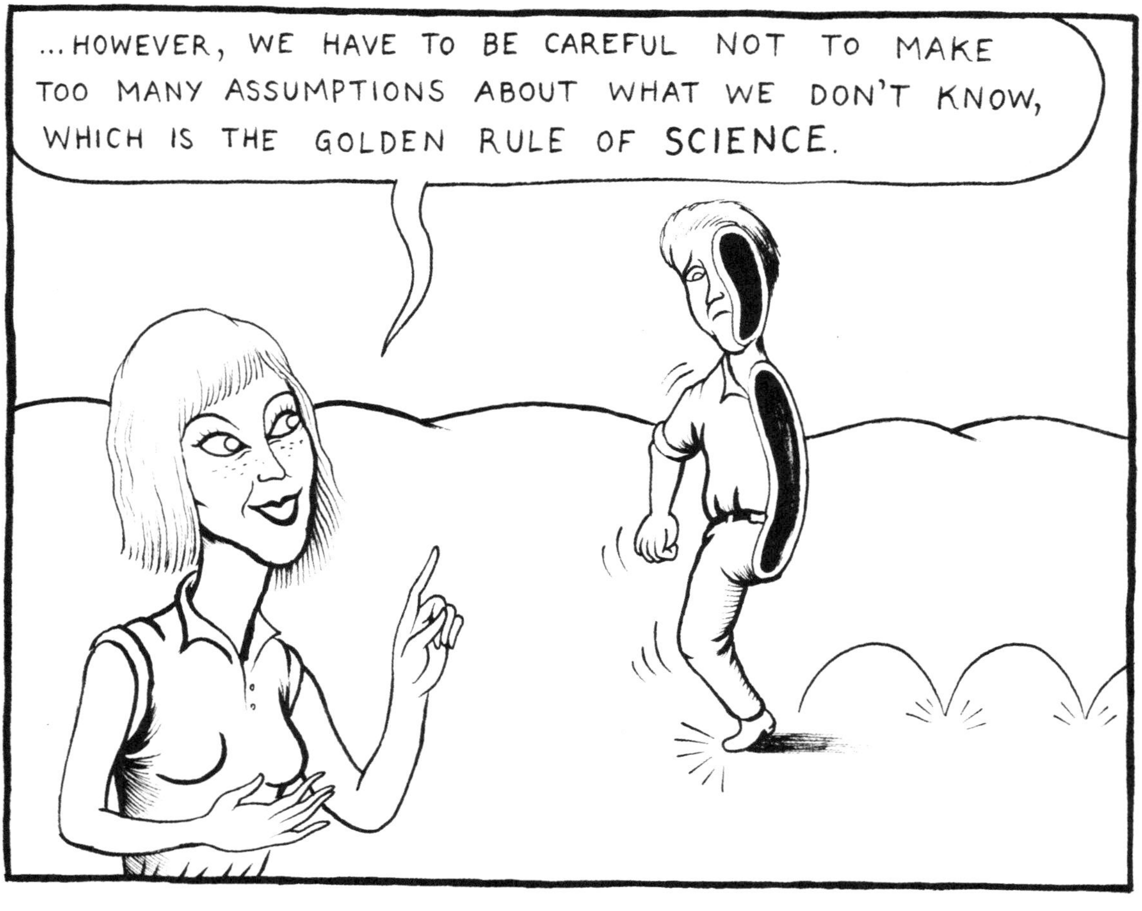
...HOWEVER, WE HAVE TO BE CAREFUL NOT TO MAKE TOO MANY ASSUMPTIONS ABOUT WHAT WE DON'T KNOW, WHICH IS THE GOLDEN RULE OF **SCIENCE**.

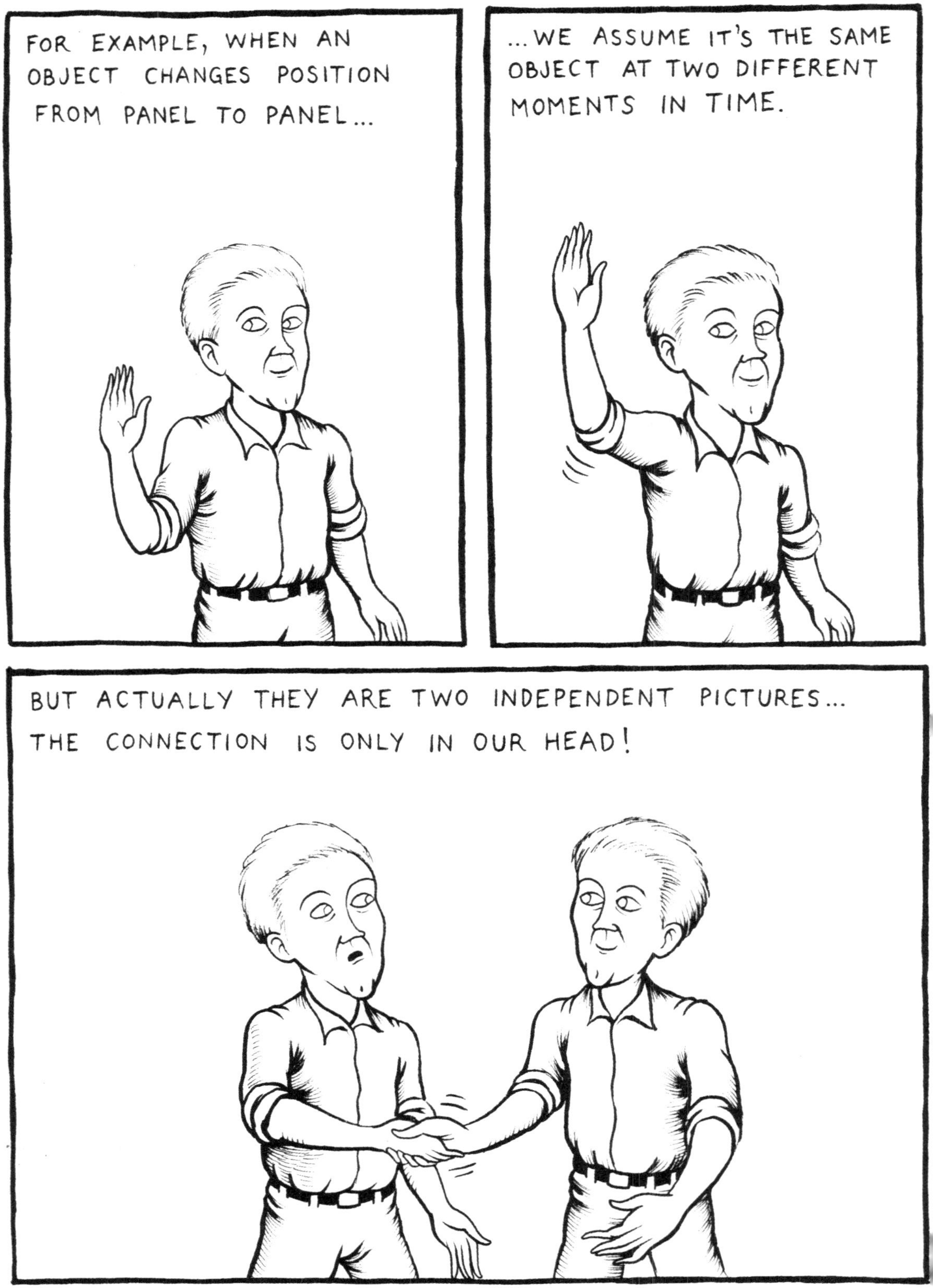
FOR EXAMPLE, WHEN AN OBJECT CHANGES POSITION FROM PANEL TO PANEL...
...WE ASSUME IT'S THE SAME OBJECT AT TWO DIFFERENT MOMENTS IN TIME.
BUT ACTUALLY THEY ARE TWO INDEPENDENT PICTURES... THE CONNECTION IS ONLY IN OUR HEAD!

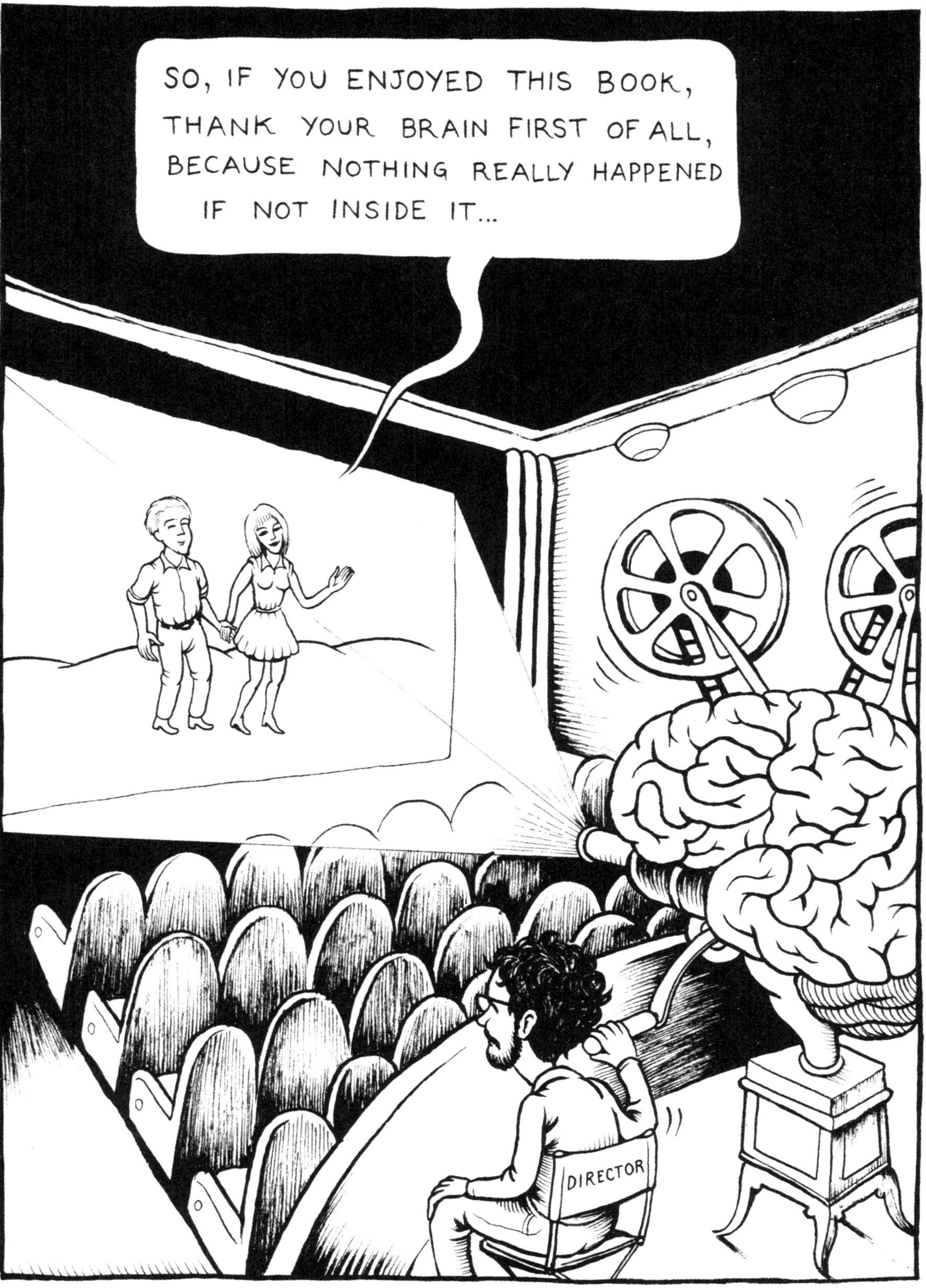
SO, IF YOU ENJOYED THIS BOOK, THANK YOUR BRAIN FIRST OF ALL, BECAUSE NOTHING REALLY HAPPENED IF NOT INSIDE IT...
DIRECTOR

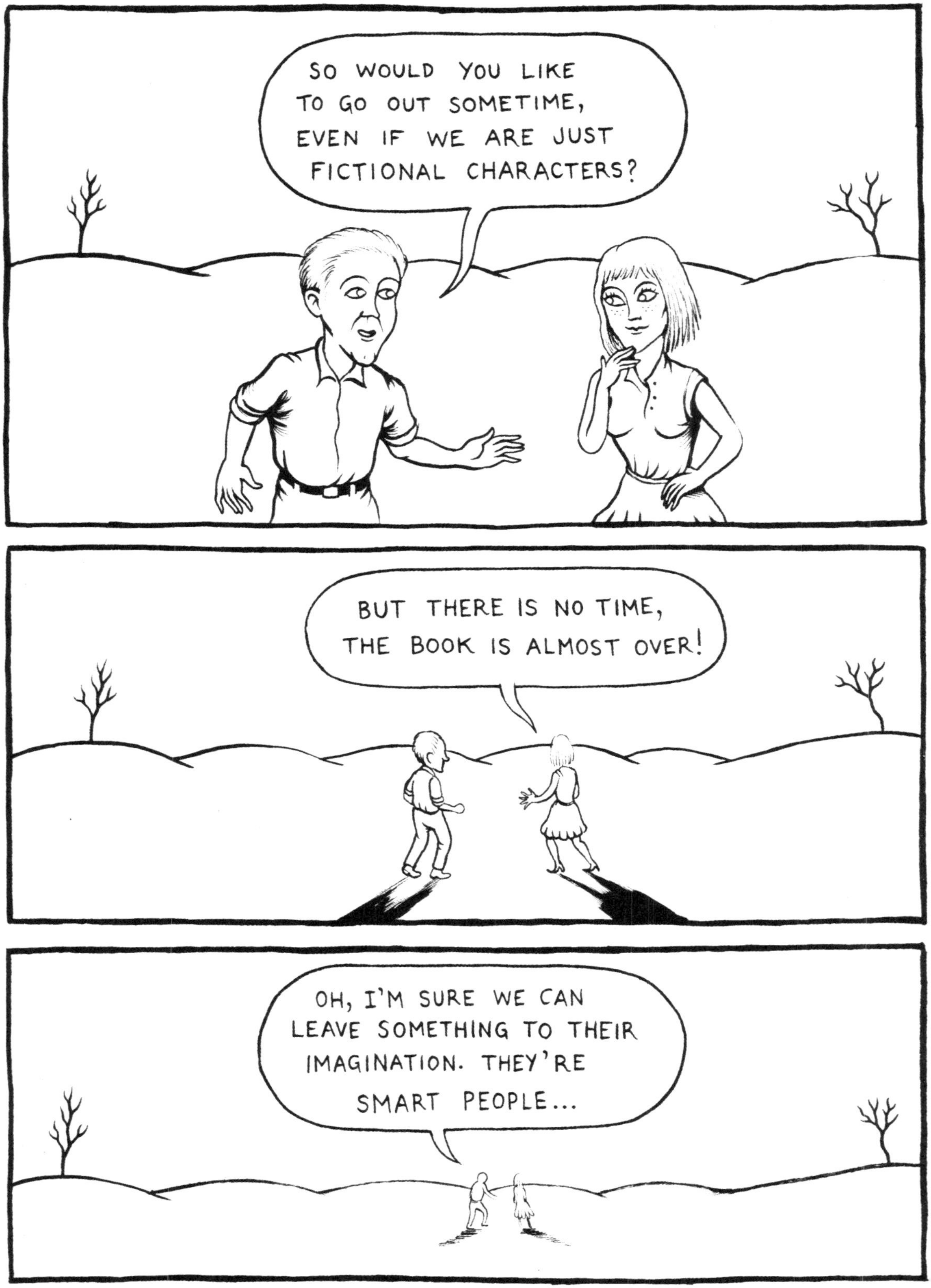
SO WOULD YOU LIKE TO GO OUT SOMETIME, EVEN IF WE ARE JUST FICTIONAL CHARACTERS?
BUT THERE IS NO TIME, THE BOOK IS ALMOST OVER!
OH, I'M SURE WE CAN LEAVE SOMETHING TO THEIR IMAGINATION. THEY'RE SMART PEOPLE...